M J Baker

HOW THE UNIVERSE OPERATES

A Metaphysical Analysis

AUSTIN MACAULEY PUBLISHERS™

LONDON • CAMBRIDGE • NEW YORK • SHARJAH

A CIP catalogue record for this title is available from the British Library.

ISBN 9781035818013 (Paperback)
ISBN 9781035818020 (ePub e-book)

www.austinmacauley.com

First Published 2023
Austin Macauley Publishers Ltd
1 Canada Square
Canary Wharf
London E14 5AA

ACKNOWLEDGEMENTS

This work is the fruit of cooperation between the author
and Dr Mark Smith who has managed the author's website
superflumina.org for the best part of twenty years. My thanks
are due, too, to Matthew Murphy for his oversight of the text
and for his suggested amendments on etymology and scientific
issues and correction of my incompetent arithmetic, to Miss Teresa
Smith for the art work on the diagrams in Chapter 4 and Miss
Elisabeth Fanning for her assistance on the layout of Chapter 5.

In honour of St Dominic, Founder of the Order of Friars Preachers, whose greatest son is the *Angelic Doctor*, St Thomas Aquinas

O Lumen Ecclesiae, Doctor veritatis;
Rosa patientiae, Ebur castitatis;
Aquam sapientiae propinasti gratis;
Praedicator gratiae nos iunge beatis.

Ora pro nobis Beate Pater Dominice,
Ut digni efficiamur promissionibus Christi.

CONTENTS

INTRODUCTION

THE ENLIGHTENMENT, THE period between the early 17th and the late 18th centuries, marked the emergence of a philosophical and theological revolution in Western thinking that had been gathering momentum for more than a century.

In philosophy it was marked by abandonment of the understanding that causation is fourfold in favour of the conception, promoted by an emergent materialism, that matter alone could explain all things. Its genesis was in the thinking of Francis Bacon and René Descartes. Newton preferred Descartes to Aristotle in his explanation of the movements of the heavenly bodies, though he retained a residual respect for the great philosopher in his insistence on the need for an efficient cause, an extrinsic agent of every effect, a view his successors were to reject with derision.

At the theological level the inchoate atheism in Luther's rejection of God's authority in favour of his own—the authority of the believer to pick and choose what he would accept or reject—developed with time. God was no longer seen as necessary in human knowing or acting. A maxim of Protagoras was adopted: *Of all things man is the measure.* Man would henceforth rule his own destiny. Religion descended into Deism, its adherents embracing the pseudo-religion of Freemasonry whose protocols were grounded in a series of oaths that mocked God to His face.

In philosophy, *materialism*, in theology, *atheism*, the one complementing the other. The 'light' of the Enlightenment was like that seen through lenses which accentuate forward vision at the expense of the peripheral. Its votaries proceeded on the simplistic understanding that the complexities of reality could be explained adequately without recourse to the views of earlier minds. They found support in the maxim attributed to the nominalist philosopher (and Catholic heretic) William of Ockham "Entities are not to be multiplied unnecessarily". The facility and faculty of distinction, based in formal causality, was lost as the need for a formal cause was abandoned. Complexities were glossed, simplistic explanations regarded as sufficient. The emerging discipline of experimental science suffered profoundly.

It is a maxim of Aristotle, one reflecting common sense, that a small mistake in the beginning becomes a big mistake in the end. (*De Caelo* I, ch. 5) The mistakes conceived during the Enlightenment have become big mistakes today. Even the best of modern thinkers succumbs to its hubris. How often does one hear, for instance, teachers of science asserting that the cause of the simple phenomenon of sphericity found in soap bubbles, in rain drops and in pellets of molten shot, is adequately explained by the disposition of surface tension, something found in their matter? It never enters their minds that these effects must have an extrinsic cause.

Modern scientists regard space as a self-existent nothing: they never stop to consider that the contention is impossible. The philosophical error spills over into logic for they treat mental being (which exists in the mind) as convertible with real being (reality). The 'thought experiments' of Einstein and his ilk are typical of the absurdity, the occasional benefits of such processes doing nothing to justify the abandonment of logical principle. One need only think of Einstein's contention that space—on any conception something utterly devoid of reality—is the cause of gravity. *Nothing* is alleged to be the cause of *something*!

Aristotle is, on any objective assessment, the greatest original thinker the world has ever produced. Henry Sire has produced an admirable summary of his achievement:

"All other thinkers have begun with a theory and sought to fit reality into it; Aristotle is the only philosopher to have begun with reality and devised a system by which to understand it. He may thus be called the only scientific philosopher, though to put it that way is to connive at the modern flattery of science. It would be equally true to say that the philosophical framework of all scientists, as of any practical thinker, is essentially Aristotelian. Aristotle took the whole of human knowledge for his study. The other ancient philosophers, other than those who were primarily scientists, ignored physics, or, as with the Epicureans, considered them only superficially. Aristotle embraced both metaphysics and physical science; and he did so in no schematic spirit but by a painstaking assessment of the scientific thought of his time. Where Aristotle accepts or rejects a scientific explanation, he does so on practical grounds, not on those of consonance or dissonance with a preconceived theory."[1]

An analysis of reality that follows the teaching of Aristotle differs fundamentally from the materialistic approach spawned in the Enlightenment. It does not deny science's findings but looks at them in a more profound light, a metaphysical light. The fundamental issue dividing the two is this: Aristotle insists that analysis shows that the greater part of reality is not material. He teaches, moreover, that the principles he enunciates apply universally. There are four causes of every effect found throughout the universe from the star *Sirius* to the computer that sits before the reader. There are no less than four; there are no more. Two of those causes are intrinsic, i.e., found in the effect, and two are extrinsic.

In contrast, materialism, uncertain about the universality of its principles, reduces Aristotle's four to one, fudges the data to account for the other three and reduces their influence to blind forces and accidents. It regards matter as evolving from one thing to another though it is quite unable to explain why the alleged developments have produced the multitude of happy results found in nature. Science's commitment to this latter-day Heracliteanism (reality in constant flux) is without support. Facts demonstrate a remarkable stability, something science is happy to take for granted in the rigour of its disciplines. That this stability contradicts the evolutionary thesis does not seem to trouble its exponents. A cavalier attitude towards strict logic is a symptom of the materialist virus.

Aristotle teaches that every material thing manifests itself in one or more of ten categories, substance and (nine different) accidents. The modern scientist, following Newton in his *Principia Mathematica,* misunderstands what is meant by substance. The metaphysician insists on the reality, reflected in the meaning of the term from the Latin participle *substans,* that a substance is 'that which stands under' one or other of its accidents, the phenomena on which the modern scientist focuses. Again, whether the thinker is considering the star *Sirius* or the computer at which he sits, this doctrine of the Categories applies. Some things, *substances*, exist in their own right, others (*accidents*) exist only in substances as, for instance, the two on which the present book focuses, light and gravity.

---

1   H.J.A. Sire, *Phoenix from the Ashes*, Kettering Ohio (Angelico Press), 2015, pp. 25-6.

Not only is the scientist constrained by this limited philosophy, he is constrained by his world view and, it must be said, his 'religion'. Regardless of what he may think about himself he is, at least inchoately, an atheist, for he engages in practical rejection of the possibility of an over-arching intellect responsible for the intricate order in the things he studies. He rejects the possibility that *this* influence has established with rigour the natures of things and conserves each in being until it dies or is corrupted. The scientist is like a man walking in a field narrowly over-taken by a golf ball who declines to investigate its trajectory to discover the agent, and the agent's intent, in favour of dissecting the ball!

Modern science lives in a sort of fantasy world, forever hinting that its exponents are about to solve the mysteries of the universe. The writings of certain of them are more pretentious than science fiction, but not as entertaining.

The reader will note that I quote Aristotle and his chief commentator, St Thomas Aquinas, as ultimate authorities. They are the doyens of metaphysics as Newton, Maxwell, Einstein and others are the doyens of experimental science. These two philosophers taught within the limitations of the experimental scientific knowledge of their age. They knew light and gravity as metaphysical accidents and their knowledge of the elements of which things are comprised was limited to a rudimentary four—earth, air, fire and water. But because they were dealing with being *simpliciter* rather than mere accidents of the material part of being (phenomena), their findings have lost nothing in importance. They insist that if we are to understand what is otherwise inexplicable about reality it must be accepted that there exists a heavenly body or *aether*. A fresh consideration of their analyses may provide us with answers to innumerable questions about the natural world.

To assist in understanding their thinking I have set forth in the Appendix to the chapter on light Aristotle's teaching in the *De Anima* (Concerning the Soul) and St Thomas's commentary on his text. I have added an occasional comment of my own. The material here is not essential to the argument and the reader may, at a first reading at least, conveniently ignore it. I have included a glossary at the end to assist the reader in understanding the terminology used.

My grasp of metaphysics came from years of study at Sydney's Aquinas Academy. The Academy did not bestow degrees and the need to earn a living as a lawyer precluded my travelling overseas to obtain philosophical qualifications.

My thinking on the metaphysical significance of *aether*, Aristotle's heavenly body, which St Thomas refers to as 'first altering body', was precipitated by reading the seminal paper of Christopher A. Decaen, *Aristotle's Aether and Contemporary Science* (*The Thomist* 68, n. 3, July 2004, pp. 375-429), and I commend anyone who wishes to plumb the topic to read what he says there before studying what I have essayed here.

MICHAEL BAKER

1. SOME METAPHYSICAL PRINCIPLES AND CONSIDERATIONS

Thinking Ontologically rather than Temporally

It is characteristic of modern science under the influence of materialism to think of reality in terms of a continuum. Thus its exponents treat the various grades of material things, from minerals through instances of vegetative, sensitive and intellective life, as parts of a process whose elements differ from each other only in their complexity—i.e., differing only *materially* from each other. Consistent with this, they look to time as measuring the process of development.

Two events may occur at the same moment yet one of them will precede the other. St Thomas Aquinas cites an example given by St Augustine that if from all eternity a foot be taken to have been imprinted in soil, the foot must necessarily pre-exist the footprint. (*Summa Contra Gentes* I, 43, 14) A practical illustration of the principle may be had by considering a boy, Patrick, chasing a ball with the Sun behind him. In the order of movement Patrick's shadow is first; in the temporal order (the order of time) Patrick and his shadow are together; in the ontological order (the order of reality), however, the boy is prior to his shadow for he can exist without his shadow but his shadow cannot exist without him. Any analysis of reality that fails to consider this order, one that fixes on the temporal alone, is defective.

It is curious that while modern science regards 'space', subsistent nothing, as breaching materialism's demands for a continuum, it does not look to some material element as filling the breach. Two influences are at war here: the first, the demand for a continuum, and the second (and more powerful), the insistence that if something cannot be detected experimentally it does not exist. Modern science's nominating of entities such as 'dark matter' and 'dark energy' in the universe show a concern to explain forces which cannot be detected by scientific instruments.

Potency and Act

Potency and act are metaphysical categories of being (and of logic) that reflect common sense. The distinction between them is easy to understand once one reduces each to the basic concept signified by its name. (Every definition should start with the *nominal* definition.) *Pot-ent-ia* (in Latin) signifies 'can-be-ness'; *act-us* signifies 'does-(be)-ness'. The man Isaac Newton is (*does be*) a scientist; the boy Patrick *can be* a scientist; Patrick's dog, Shep, *cannot be* a scientist. Even though it is as yet undeveloped, there is a reality in Patrick that is not in his dog. Newton has the habit of science *in act*; Patrick has the habit of science *in potency*. Thus potency is something real, not imaginary.

Water cold is *in potency* to be hot water; it can be brought from cold to hot by something which is hot *in act* (the Sun; a fire). The countryside on a dark night is *in potency* to be illuminated; it can be brought from invisible to visible by something that emits light, i.e., something that has light *in act* (the Sun; a searchlight). Matter (prime matter) is in potency to be a rock, a tree, a dog or a man. All that is needed is something in act with respect to each of these kinds of being to render it so. The principle underlying the distinction is that nothing can be brought from potency to act (under any respect) save by something which itself is already in act (in that respect).

The Doctrine of Causality

The scientist sitting at his computer will acknowledge that the machine in front of him is comprised of matter in various sub-categories, metals, glass, plastic; elements, or compounds of various of the 118 elements, which the metaphysician calls 'secondary matter', matter already bearing some formality. Without this substrate there would be no computer. The scientist will also acknowledge that these instances of matter in its sub-categories must be ordered in a certain critical manner else he will not have a computer at all, but something less subtle. The influence that determines the elements is the artificial *form* that makes the thing be a computer rather than a television set, a radio, or a device for mixing cake ingredients. The metaphysician refers to these two influences as the *material* and the *formal* causes of the computer.

But more is required before the computer can be an existing thing. First, there has to be a maker, or makers (referred to as the *efficient* cause) without whom the computer could not come into existence. And secondly, and most critically, there has to be a *final* cause, the reason for the computer being conceived in the first place as a device to aid the scientist in his work. This cause begins in the mind of the one who conceives it and is realised in the device produced. It is the first cause, and the last! Two causes are *intrinsic*; they remain in the computer. The other two are *extrinsic*, outside the computer and, once it is produced, are no longer essential to its continuance or its operation.

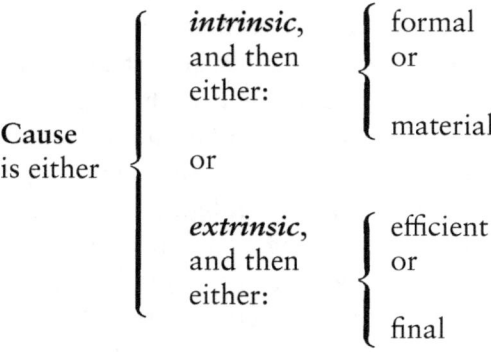

Form causes *by determining;* matter causes by *being determined*. A little reflection makes this plain. The matter that goes to make up a computer could constitute any of a number of things. The illustration used assumes matter at a certain degree of refinement (secondary matter). But matter simply amorphous, matter taken as the 'stuff' underlying the elements and their compounds (prime matter), is quite indifferent. It can be anything—air, water, marble, a tree, a man. The *formal* cause, in contrast, is what makes matter be *this* particular thing; air, water, marble, a tree, a man.

Now, art imitates nature. As with artificial things, every *natural* thing has four causes, formal and material, efficient and final. The *formal* and *material* are easy to see. Recognition of the *efficient* and the *final* causes of natural things brings in other considerations.

Just as every *thing* (whether natural or artificial) has four causes so, too, does every *action* we perform. Consider the father pushing his little son in a toy car. The *matter* of the father's act (the one undergoing it) is his son in the car; the *form* is the accident of *action* (in the father) which is felt as *passion* in child and car; the *efficient* cause is the father; the *final* cause is the desire of the child transmitted to, and accepted by, his father to feel the pleasure of being propelled.

This doctrine of fourfold causality applies universally, to the scientist at his computer and to the star *Sirius*, the brightest star in the sky some 8.6 light years distant from us.

Calculation and Causation

Notwithstanding that he favoured the views of Descartes over those of Aristotle, Sir Isaac Newton had not forsaken the rational demand taught by Aristotle that there must be an extrinsic cause of every effect. He understood that his calculation of the movements of the heavenly bodies was not to be identified with their causation. Newton wrote to the Rev. Richard Bentley the following:

> "Gravity must be caused by an agent [acting constantly] according to certain laws, but whether this agent be material or immaterial is a question I have left to the consideration of my readers."[2]

The confusion of these two categories, *calculation* and *causation*, is one of the manifold effects of the errors instituted by Descartes in mixing the subjective with the objective.

Scientists are preoccupied with *calculation* in respect of the subjects of their investigations and they carry out the associated tasks with great accuracy. But it does not follow that thereby they expose their causes. Indeed, it is impossible for them to do so, for observations and measurements do not penetrate further than the surface of the things they investigate. The limitations of their discipline and the dominating influence of materialism excludes the possibility of a more profound analysis.

Contrary to the accepted view about him, Einstein did not expose the causes of the motions of the universe—which he attributed to some element intrinsic to empty 'space' (for heaven's sake!)—he only made their *calculation* more precise.

---

2   Cf. http://www.newtonproject.ox.ac.uk/view/texts/normalized/THEM00258

The Categories of Being

If the reader is to understand the arguments proposed here it is necessary that he have some grasp of the metaphysical doctrine of the *Categories of Being*.

Some things 'be'; other things 'be-long'. One never sees the colour blue, for instance, except in some thing, like the sky, the sea or a painted surface. 'Up-side-down' is a reality never found by itself, only in a being such as a child on parallel bars or an aeroplane performing aerobatics. 'When' and 'where' can only be said about some objective thing; they have no independent existence. There is no such thing as 'posture' in the absence of something posing; no such thing as a grin without the Cheshire cat!

Every substance has nine *accidents*: the first, *quantity*, provides physical extension and parts, i.e., provides the substance with a body.[3] The next accident, *quality*, makes the body be of *what kind* (*qualis*) giving it density, permeability or impermeability, hardness or softness, texture, colour, heat, and so on. The remaining accidents *relation, when, where, action, passion, habitus* and *situs,* determine substance in every other possible fashion. One of St Thomas's commentators puts it this way:

"Among all... accidents it is proper to *quality* to render the subject formed and qualified... because *quality* among all the accidents properly ennobles and qualifies the subject. For *quantity* quantifies and rather materialises its subject by extending it and ordering its material parts. The remaining [accidents] either order their subject towards another, as does *relation*, or depend upon something extrinsic ordering it, as do the last six. What... essential difference does essentially, *quality* does accidentally, namely to form and qualify what is potential and formless..."[4]

Here are the ten classes in order:

---

3   In his text, *God and the Theory of Everything*, (2012), Dr D. G. Boland of Sydney's *Centre for Thomistic Studies*, points out that Descartes rejected the metaphysical understanding of *substance* and substituted for it first accident, *quantity*. The text may be viewed at http://www.superflumina.org/PDF_files/d-boland-god-the-theory-of-everything. pdf.

4   John of St Thomas, *Cursus Phil.*, I, p. 609b, quoted in *Ostensive Metaphysics, Treatise One, Ontology*, A. M. Woodbury Ph. D, S.T.D., a text of Sydney's Aquinas Academy, n. 1161.

Substance   a thing which exists in itself
Accidents (which exist only in some thing)
>   *quantity*—gives substance spatial extension and parts
>   *quality*—renders it *of such sort*, in a variety of ways
>   *relation*—orders it towards other things
>   *when*—at this time
>   *where*—in this place
>   *action*—renders it apt to produce change in another
>   *passion* – whereby it suffers some change from another
>   *habitus*—whether it is clothed or covered
>   *situs*—posture, e.g., upright, lying, sitting etc.

Socrates is a substance: that he is a teacher to his pupils is an accident—*relation*. A dog is a substance. That it has a body with differing parts is an effect of its proper *quantity*. That it is brown is a *quality*. That it is chasing another dog belongs to *action*. That it wears a collar is of the accident *habitus*; that it is standing, of the accident *situs*. Water is a substance; its presence in a pond is an accident (*where*). In short—

> **substance = be-in-self**
> **accident = be-in-other.**

A note of caution: the elaboration of the reality signified by the word *substance* here is *not* how Newton, influenced by Descartes's teaching, conceived of it in the *Principia Mathematica*. Newton's definition reduced it to what the metaphysician refers to as *quantity*. This loss of understanding of what is meant by substance brought with it a loss of the sense of the need to distinguish substance from accidents. Einstein laboured under the same disabilities. This led to him attribute to accidental realities, or imaginary realities (i.e., beings whose only reality was in the collective minds of scientists), burdens which could only be borne by a substance, i.e., by a thing in itself.

The Doctrine of Analogy

This part of metaphysics has to do with the way we use terminology: it is strictly a part of logic. A term (known in logic as a predicate) may either be *univocal*, *equivocal* or *analogous*. A *univocal* term applied to

17

two different objects signifies a character in each which is simply the same, as the term 'animal' when said of a dog, of a cat and of a bird signifies in each the same reality.

An *equivocal* term applied to two different objects signifies something in each which is simply un-same, as the term 'cricket', when said of an insect of the order *orthoptera* and when said of the game played with a ball and a bat, signify completely different realities; or as the term 'board' when said of piece of timber and a collection of persons involved in the management of a company signify completely different realties. The only thing in common is the term, the name.

An *analogous* term signifies something in different objects which is both same and un-same—and more un-same than same! For example, the term *healthy* may be said of a climate, of a type of food, of physical exercise or of a living body with a normally functioning organic constitution. It is said properly only of the last of these, for a normally functioning organic constitution is of the essence of health. It is said of the other three by extension, or attribution, for climate is a cause of health, and a type of food may dispose to health, as may physical exercise. This is *analogy of attribution or proportion*.

But there is another way of using analogy. The predicate 'good' may be said of a dog and of a man. Both are said properly, indicating something in each which is suitable to appetite; but the first signifies the limited attractiveness proper to a creature acting according to its nature, while the second signifies the much greater reality of the attractiveness of an intellectual being living in accordance with the moral law. In the same way we use the predicate 'being' of a creature and of God. The creature is properly called a being because it has 'be', or existence; it exists, if contingently. But 'being' is said pre-eminently of God who *is* 'be', or existence; who exists necessarily. This species of analogy is called *analogy of proportionality*. There are other subtleties to the doctrine which need not concern us here.

These distinctions are of great importance in an age where, because they fail to distinguish between *univocal* and *analogous* terms, men often argue at cross purposes. Because the philosophy to which they subscribe disposes them to confuse realities which are distinct, scientists are prone to such errors.

Moreover, materialists tend to confuse what exists in the real (real being) with what exists only in mind (mental being) and pass from the one category to the other without discriminating. Because they can *imagine* or *conceive of* some development, they think it must exist in the real. This is the chief defect in much of what passes for the reasoning of evolutionists: they cannot prove the alleged change from one type of creature to another *but they can imagine it*. This suffices for them to claim their thesis is proven. The defect colours the workings of the device much loved of theoretical scientists and mathematicians, the 'thought experiment'. It is responsible for the assertion that Einstein's calculations of the behaviour of the accidents of energy, mass and the speed of light, and assertions of physical dimensions additional to the three dimensions, represent reality, when they are no more than predictions (albeit accurate) of how things will behave. They are but conceptual devices illustrating their formulae. *Calculation is not to be identified with causation.*

Contrary and contradictory

It is customary in the modern world to use these two terms interchangeably but they signify different realities.

Contrary opposition is opposition between two classes which are furthest removed from each other among those which belong to the same genus as, for instance, red and blue (in the genus of colour), pious and impious (in the genus of religious inclination), kind and cruel (in the genus of moral conduct).

Contradictory opposition is the opposition between a term and its negation, as between man and not-man, between white and not-white.[5] Here the two terms are not merely mutually exclusive but they are exhaustive of all possible things. When I was teaching philosophy I advised my pupils that they could adequately divide the whole of reality into Ambrose (one of the students) and not-Ambrose. This was division using contradictory opposition.

The distinction between the two should be noted when reading what St Thomas has to say on light below.

5   Cf. G.H. Joyce, *Principles of Logic*, pp. 36-7.

Materialism's Impossible Premise

Modern science treats the space beyond Earth's atmosphere as non-being-somehow-existing. It does the same at the atomic level when it asserts with confidence that a material body is comprised, as to 99%, of 'empty space'. It is as characteristic of the philosophy of materialism to be ignorant of the ontological order—of the distinction between something and nothing—as it is to be insouciant about causes (or any cause other than emollient matter). I address this threshold problem in the first part of the text below.

The influence that gives life to a living thing, the soul, is not detectable scientifically. Science regards it as some sort of emanation of matter but can give no explanation as to how it arises or in what it consists; but no scientist would assert that the soul is not some thing, not some reality. Now, in similar fashion, simply because the 'space' beyond Earth's atmosphere and that which makes up 99% of a material body cannot be detected scientifically it does not follow that it is not some reality.

When we use the term 'nothing' we are using the mind to give a positive value to something negative: a lack, an absence, of being. We do the same thing when we use the term 'night'. We give a positive value to something negative, the lack of light that follows on the setting of the Sun. The distinction is that between what exists only in the mind, and what exists not only in mind but also in the real. 'Nothing' exists only in the mind, as 'night' exists only in the mind, for the reality each term signifies is not something; it is *the lack of* something.

It is impossible that 'space' or 'void' be comprised of non-being-somehow-existing because 'nothing', or non-being, cannot exist in the real; it can only exist in our minds.

St Thomas exposes a further argument in his discussion of the senses in Lecture XIV nn. 6, 20 of his Commentary on Aristotle's *De Anima* Book II, Chapter 7. Every sense, he shows, requires a certain touch. This applies to the sense of sight no less than for the other four senses. In the senses which operate at a distance from their object (hearing and sight) there must be a medium, a material continuum, between the organ and the object. If we are able to see the light from a star it can be only because there is an uninterrupted material medium between the eye and the star. It is impossible therefore that there could exist a breach of this continuum. Moreover, this (transparent) body must be undetectable. If it were otherwise, it would impede the vision of what it conveyed, a facility remarked in passing by Christian Huygens in 1678.

"I do not find that anyone has yet given a probable explanation of the first and most notable phenomena of light, namely, why it is not propagated except in straight lines, and how visible rays, coming from an infinitude of diverse places, cross one another without hindering one another in any way... (*Treatise on Light*, Ch. 1)

Hence, wherever science is unable to detect the existence of anything—where it asserts 'empty space' or 'void'—there must be an existing something. The crucial question is just what this something is.

Materialism and Atheism

The rise of atheism—belief in no-God—is in large measure a function of the dominance in modern society of materialism. The two have the same provenance and the same insouciance about causes, especially ultimate causes. They complement each other.

It may come as a shock to the reader of this book to be told that his adherence, whether implicit or explicit, to the atheistic belief system will prevent him understanding the metaphysical reality of how the universe operates. But it will. The reason is that such understanding demands that there be admitted an overarching intellect involved in the universe's final and efficient causality.

But all is not lost. The reader should consider the case of the American philosopher Mortimer Jerome Adler (1902-2001), self-confessed pagan, who could not resist the force of the reasoning of Aristotle and St Thomas demanding he acknowledge the need for efficient and final causes of reality. The reader might dip into Adler's work *How to Think about God: A Guide for the 20th Century Pagan* (New York, Simon & Schuster, 1980: [ISBN 0-02-072020-3]) available on Kindle.

2. SCIENCE AND ARISTOTLE'S *AETHER*

MODERN SCIENCE MAINTAINS that, apart from the stars, planets, moons, asteroids and various atoms found in it, there is nothing in interstellar space. Consistent with this position science has long maintained that light waves and electromagnetic radiation do not require a medium in which to travel.

I asked a friend of mine qualified in science:

- Why, if there is nothing in interstellar space, does not this 'nothing' present an absolute barrier preventing sunlight, moonlight or light from the stars reaching the Earth?

- How can something—light waves (or particles) or electromagnetic radiation (or planets, or asteroids, or stars, for that matter)—pass through this 'nothing'?

Anticipating the answer he gave to the latter question—"There is nothing to impede them"—I asked: Why, then, is the speed of light, c, determinate (299,792 kilometres per second)? Why is it not infinite? He did not know: modern science does not know; nor, consistent with the head-in-the-sand mentality towards anything it cannot explain, does the philosophy on which modern science is founded concern itself over the issue. It simply takes the fixity of c for granted.

The concept of ether as a medium was dismissed, my scientific friend told me, early in the 20th century largely as a result of Einstein's work and following the celebrated Michelson-Morley experiment in 1887. Even in deep interstellar space there are millions of atoms per cubic metre. He conceded that the space between those atoms was no less puzzling than is the space between the nucleus and electron shell of every atom of every element. If this space were removed, he said, our planet would be reduced to the size of an orange. Experiments show that most of what we call 'matter' is nearly all free space: and science cannot explain why the speed of light is limited.

Materialism maintains that there is no reality in anything not material; by which is meant anything that cannot be measured physically. In the year 1500 there was hardly a materialist in the world. By the year 2000, there was hardly a thinker who was not a materialist. I have set out elsewhere the history of the development (or, as I contend, the decline) in thought which led via the systematic denial of Aristotle's doctrine of causality to the acceptance of materialism's banal imperatives.[6] The cause, I contend, was a religious one, the rise and flourishing of Protestantism which, despite its protestations of religion, is inchoately atheistic. Atheism cannot flourish unless philosophy is deprived of its ground in reality and Protestantism provided the catalyst for the necessary dumbing-down of thought. The burgeoning of what was begun in the sixteenth century occurred in the two centuries that followed, the period known as the Enlightenment.

The scientific revolution is generally dated from 1543, the year of publication by Nicolaus Copernicus of his *De Revolutionibus Orbium Caelestium* and by Andreas Vesalius of his *De Humani Corporis Fabrica*. It began, thus, just as Protestantism and its atheistic tendency was taking root. When, in June 1661, the young Isaac Newton entered Trinity College Cambridge, the University followed the teachings of Aristotle in natural philosophy. Newton preferred the thought of Descartes, as he preferred the observations and inductive reasonings of astronomers Galileo, Copernicus and Kepler. He borrowed Aristotle's notion of *aether* as necessary to transmit forces between particles (among which he counted light) but, because he rejected Aristotle's metaphysics, he misunderstood his teaching. Under the influence of Descartes' mechanistic views he treated ether as a greatly rarefied instance of common matter. He was to pass this misunderstanding to his successors.

In the 1860s James Clerk Maxwell established that light was a species of electromagnetic radiation and, using the data then available, he determined its speed in a vacuum at 310,740 km/s. He wrote:

> "The agreement of the results seems to show that light and magnetism are affections of the same substance, and that light

---

6   Cf. *Pity the Poor Atheist*, at https://www.superflumina.org/PDF_files/pity_theatheist.pdf

is an electromagnetic disturbance propagated through the field according to electromagnetic laws."[7]

Science has now established that all electromagnetic radiation travels in a vacuum at a determinate speed, 299,792.458 km/s, remarkably close to Maxwell's figure. As had Newton before him, Maxwell postulated the necessity of a luminiferous ether to carry these waves but the ether he assumed was, like Newton's, a rarefied common matter.

The experiment Albert Michelson conducted with the assistance of Edward Morley at what is now Case Western Reserve University in Cleveland, Ohio, in 1887 was designed to detect this postulated medium. They reasoned that, however 'ethereal' it might be, ether must have mass and inertia. By means of an ingenious device Michelson had invented (the 'interferometer') they took a source of white light and split it into two beams travelling at right angles to each other out to mirrors at a distance which returned the beams to a common detector. Any slight difference in the time the two spent in transit was detectable via the phenomenon known as *interference* where the combining of two sets of light waves slightly out of phase will manifest itself in a new (combined) wave pattern. They discovered no pattern which was not explicable by experimental error. The speed at which light travelled was the same for all observers whether in the direction of travel or at right angles to it. There was no evidence that ether possessed mass or inertia.[8] Following their materialist protocols they reasoned, 'if it is not detectable, how can it be said to exist'?

Aristotle's Aether

In this summary of Aristotle's teaching on *aether* I have drawn on the paper by the American philosopher, Christopher A Decaen, referred

---

7   In his paper *A Dynamical Theory of the Electromagnetic Field*, 1864

8   The experiment has since been repeated on many occasions with greater precision and the same results. Cf. http://en.wikipedia.org/wiki/Michelson%E2%80%93Morley_experiment

to in the Introduction.[9] Decaen uses the terminology of metaphysics, foreign to minds trained in the simplistic categories of materialism. But the concepts of metaphysics are no less understandable than, in their disciplines, are those of the Special and General theories of Relativity and the theory of Quantum Mechanics. I have endeavoured to assist the reader with a glossary and with footnotes. Although St Thomas differed from Aristotle on a number of topics, on this one his mind followed that of Aristotle closely.

Decaen demonstrates why Aristotle saw the circular movement of the heavens as significative of a radical difference between the mundane and heavenly bodies.

"[T]he principal datum of nature that [Aristotle] wishes to explain with *aether* can be experienced firsthand by spending the night under the stars and watching their motion as the night hours pass. One finds himself at the center of a perfectly circular pilgrimage of stars traveling from east to west, as though each of the heavenly bodies [was] embedded on a dark orb revolving around the Earth. This nightly, and a related yearly, uniform circular motion of the stars should provoke a question: Why should this apparently natural motion occur in the sky, indeed in most of the cosmos, but not here below, where few things seem to move in circles without being coerced? This peculiarity [of circular movement] is all the more striking when one notices that these same heavenly bodies and their motions are never seen to change, much less corrupt or cease... This appearance of eternity and incorruptibility is strengthened by the astronomical records... 'For in all time gone by, according to all records handed on from one [generation] to the next, no change has ever appeared either in the whole of the containing heaven or in any proper part of it.'"

Reasoning that a void, a region not filled by a material substance, is physically impossible Aristotle concluded that the heavens, the vast expanse between the visible heavenly bodies and the world in which we live, must be filled with an invisible material medium. Decaen again:

---

9   *Aristotle's Aether and Contemporary Science* is divided into three parts: 1. *Aristotle's Celestial Substance*, where the author details Aristotle's teaching; 2. *The Fate of Aether in Classical Physics and the Special Theory of Relativity*, where he exposes the misunderstanding of Aristotle's concept by Newton and his successors; and, 3. *Contemporary Science's Resuscitation of Aether*, where he shows science's return to a sense of *aether* as essential to the theories of Relativity and of Quantum Electrodynamics.

"Not only are the *stars and planets* made of a different kind of substance, but—given that such perfect transparency is present in something that manifests no signs of ordinary matter's downward or upward tendency, but either is perfectly yielding to the visible circular motion of the stars and planets, or moves with them—so must be the subtle matter surrounding them. Thus, Aristotle applies the name 'aether', or more frequently, 'the first body', to whatever fills the volume of space between the Moon and the outermost sphere of the fixed stars. It *is* itself 'the heaven... the continuous body in the place after the outermost circumference of the whole, in which are the Moon, and the Sun, and some of the stars [i.e., the planets].'"[10]

The modern scientist may mock Aristotle for relying on appearances which he can demonstrate to be illusory. But a moment's thought will show that the appearances are not so illusory after all. What is the apparent circular movement of the stars but a function of the rotation of the Earth?—a circular movement. The Earth rotates around its axis and it revolves—in a circle—around the Sun. The other planets do the same. Earth's satellite, the Moon, also moves in a circle, around the Earth, and the moons of other planets do the same. The far galaxies demonstrate a circular pattern in the layout of their constituent stars. Why, then, should we mock Aristotle for ascribing circular movement as a property of his postulated heavenly body? There is, moreover, other circular movement in the cosmos almost infinite in extent in which *aether* seems intimately to be involved, that of electrons about the nucleus of every atom.

Although it is material, *aether*, the heavenly body, does not share a mode of being with other things. Matter and form in *aether* are not predicated univocally with matter and form in other material substances, but analogically.[11] In other words *aether* does not share the physical attributes of other material things. St Thomas says that matter and form in *aether* are so perfectly united that the one exhausts completely the potency of the other. *Aether* lacks the tendency to become something else (the principle of corruption) and is as incapable of being generated as it is of being corrupted; incapable of growth or alteration. *Aether* is simple; it is immutable, not subject to change in substance, quantity

---

10   Christopher A Decaen, *Aristotle's Aether and Contemporary Science,* op. cit., section 1 A.

11   Using analogy of proportionality. See Chapter 1 for an explanation.

or quality—though apparently so in respect of place. It has no weight or lightness. It is not susceptible to temperature or pressure. *Aether* is intangible, enjoying, as Decaen remarks, the paradoxical characteristic that:

> "being wholly impervious to alteration entails... [it] be perfectly pervious to something trying to press upon it."

The scientist will doubtless contend that, since they are not verifiable experimentally, these properties are nothing more than assertions. But he will reach that conclusion not because he is a scientist, but because he is a *materialist*. If reason requires that we posit the existence of some thing, it is no answer to say that it is not detectable experimentally. Nor does it justify rejecting properties which reason may conclude the thing possesses. The 'black holes' and 'curved space' posited by Einstein's General Theory of Relativity, the 'dark matter' and 'dark energy' postulated by scientists in more recent times are, none of them, detectable experimentally. That is no reason for denying their existence if discernible effects justify no other hypotheses. It was precisely from discernible effects that Aristotle posited—and St Thomas endorsed—the existence of *aether*.

There seem to be a number of properties of this remarkable substance, attributes which go far to explain issues that modern science has so far been unable to resolve.

The Properties of Aether

TRANSPARENCY

The first property is transparency. Aristotle regarded this as a positive nature and science seems, implicitly, to agree that transparency is not merely a privation. As Decaen says: "if darkness is the privation of light and colour, transparency cannot be."[12] Both Aristotle and St Thomas understood light to be the "act of the transparent forasmuch as it is transparent". Consistent with this, *aether* is the substance which

---

12   Christopher A Decaen, *Aristotle's Aether and Contemporary Science*, op. cit., footnote 48.

universally is in potency to illumination. Decaen concludes to the existence of this quasi-sensible quality:

> "[I]f we consider that nothing around us is perfectly transparent—one can see only so far even through air—and that the distance between the Earth and the stars is almost inconceivable, one sees that *aether* must be the most perfectly transparent substance in the cosmos."

St Thomas suggested that all other bodies are called 'transparent' only by participation in the nature of *aether* just as things are called 'hot' by participation in the nature of fire.[13] Aristotle has this to say:

> "Neither air nor water is transparent because it is air or water. Each is transparent because there is contained in it a certain quality which is the same in both and is also found in the eternal upper body."[14]

Modern science may provide a better explanation with its understanding that the atomic structure of every material thing is largely comprised of 'space'. This vacuum at the atomic level presupposes, just as much as does that of inter-stellar 'space', the presence of *aether*. Thus, the ability of transparent bodies (air, water, glass, etc.) to permit the passage of light may be explained by the fact that their atomic structures do not impede (or better, do not completely impede) the *aether* in their interstices from its proper operation. Accordingly, on this analysis *aether* is not simply a substance with supreme transparency it is universally the substance that permits the passage of light.[15] *Aether* is *the transparent*.

---

13   I rely here on Decaen's citation of St Thomas in *In II Sententiae* d. 13, q. 1, a. 4; *Summa Theologiae* I, q. 67, a. 3; *In II De Anima*, lect. 14. n. 22; and *De Sensu*, lect. 6, nn. 7-9. The quote from the *Summa* does not go this far. I have been unable to check the other sources.

14   *De Anima*, Bk. 2, Pt. 7. The critical word here in the Greek is *phusis* which means 'nature', from which we get 'physics'. One translation has it as 'substance' but this is inaccurate. I have translated it as 'quality'. In a personal communication to the author, Dr Decaen puts it in this way. "I think the eternal upper body Aristotle is speaking about IS the aether... I think Aristotle is... saying that air and water are transparent because they participate (less perfectly) in the nature of the aether itself, which (in this context) is simply perfect transparency..."

15   This is not a view with which metaphysicians would necessarily agree. Neither Dr Decaen nor Dr Don Boland agrees with my analysis.

Without it there would be no propagation of light. Without it we could not see the page in front of us.[16]

NON-RECIPROCAL AGENCY

The second positive property might be termed 'non-reciprocal agency'. It was clear to Aristotle, as it was to St Thomas, that in the coming and going of the seasons, in the tides and in other ways, the heavenly substance which included Sun, Moon and stars, affected the world below. Yet there was no evident reciprocity of effect. Aristotle concluded that *aether* affects ordinary matter but is not affected by it in return.[17]

> "While usually the thing touching is touched by what it touches... still it also occurs... that only the mover may touch the moved, while the thing touched does not touch the one touching it...[18]

And St Thomas in his commentary:

> "Bodies act upon each other by touching... But this should be understood [only] when there is mutual contact as happens in those things that share a common matter... The heavenly bodies, however, because they do not share a common matter with inferior bodies, act upon them such that they are not acted upon by them; they touch and are not touched."[19]

Their analyses may seem to be grounded on a number of false premises. The apparent lack of reciprocity between the heavenly and mundane bodies might be explained by the immensity of the distances between them and by gravitational forces, of both of which issues the two philosophers were ignorant. We know, moreover, from science's discoveries that the heavenly bodies, the stars, planets, moons and other asteroids

16 And this is not the half of it. Without *aether* the very atoms of matter could not exist. One need not even begin to think about the interaction of the heavenly bodies.

17 There is not room here to show that Aristotle's analysis is not necessarily contradicted by current cosmology which would isolate Sun, Moon and stars from the hypothesised aethereal matter. The reader should study Decaen's paper.

18 *De Generatione et Corruptione*, Bk 1, Pt. 6; and cf. footnote 50 in Decaen, *Aristotle's Aether and Contemporary Science*, op. cit.

19 *Commentary on the Physics of Aristotle*, Bk. 3, Lect. 4, n. 5; and cf. footnote 51 in Decaen, op. cit. Apparently St Thomas did not comment on Bk. 1, Pt. 6 of Aristotle's *De Generatione et Corruptione*.

are comprised of elements of matter also found on Earth, so that St Thomas's conclusion might be said to be disproved. Due to the limitations of the science at their disposal both philosophers were unaware of these facts, and their identification of the heavenly bodies with the *aether* which was their matrix is understandable. But their ignorance of such matters does not destroy the force of their comments about *aether* itself.

Decaen remarks that their teaching seems to contradict the Newtonian assertion of equal and opposite reaction among bodies. But the apparent breach of Newton's law can be resolved if *aether* is understood as being of a different, and superior, order of materiality to that of common bodies. Acknowledging this characteristic in *aether* of *exercising force without reciprocal effect* may, as we will see, explain issues that have troubled science for three hundred years.

IN NO PLACE

There is a third positive property, albeit negatively expressed. *Aether* does not, simply speaking, have location. *Place* is one of the nine predicaments (accidents) of every body. Aristotle defines it as "first immovable surface of circumambient body".[20] But *aether*, the substance which, on this contention, fills the cosmos from the level of the atom to the periphery of the solar system, has no container. Rather, *aether* is the container of everything else. It establishes *place* for everything else.

SOURCE OF GRAVITATIONAL FORCE

Newton formulated his universal law of gravitation as directly proportional to the product of the masses of the relevant bodies and inversely proportional to the square of the distance between them. Yet he did not consider gravity, as his successors regard it, as a force of attraction. He considered it one of repulsion. Nor did he regard gravity as essential and inherent to matter. He attributed it to a discrete, independent, particle he called a *fluxion*. He rejected the understanding of interstellar space as a great vacuum, regarding it as filled with the fluent matter he had postulated. In correspondence with Richard Bentley, Master of Trinity College, he said this:

20   *Physics* IV, 5; (212 a, 22)

"That gravity should be innate, inherent and essential to matter so that one body may act upon another at a distance through a vacuum without the mediation of anything else, by and through which their action and force may be conveyed from one to another, is to me so great an absurdity that, I believe, no man who has in philosophic matters a competent faculty of thinking could ever fall into it."[21]

Now, if *aether* is the intangible sea in which all matter subsists it touches all matter. *Prima facie* then it provides the mediation Newton required. But there is a problem. If *aether* touches the heavenly bodies but is not touched by them in turn, it cannot be the medium of their mutual influence. What, then, is the source of gravitational force? Surely Newton was right when he said that there is nothing in a heavenly body such as the Sun, Earth or Moon which requires that it exercise attraction on another.

My scientific friend said that our apprehension of certain of the conclusions of Einstein was 'counter intuitive'. In other words, the findings which Einstein had postulated (e.g., 'curved space') were opposed to the natural inclination of the mind. Is there something similar here? What if, notwithstanding that gravitational force is predictable and measurable, and apparently a function of the mass of the bodies involved, it is generated not by the bodies themselves *but by the aether in which they subsist*? This conclusion would confirm Aristotle's assertion of *aether's* non reciprocal agency and support Newton's thesis, if not his explanation. On this hypothesis, if we assume for the purposes of argument that Newton's Universal Law of Gravitation is valid for all cases, it would demonstrate this element of *aether's* agency with scientific precision.

Decaen provides this synopsis of recent discoveries in support of the proposition.

"According to accepted theory, the expansion of the universe should be decelerating due to the gravitational drag of massive

---

21   The letters to Dr Bentley are, according to one internet source, dated 10 December 1692, 17 January, 11 February and 14 March 1693. I have been unable to locate the date of the relevant letter. The passage is apparently reproduced in an essay, *General Scholium*, appended to the Second (or Third?) Edition of his *Principia*. I have not checked the source. Cf. Wikipedia sub cap. *Newton's law of universal gravitation*.

bodies such as planets and stars. However, observations on a number of distant supernovae over the past ten years are suggesting that some hitherto unknown repulsive force from an unknown energy source is accelerating the expansion. And worse, this force does not appear to be coming from one region of the universe; rather, it appears to be coming from all directions, or more specifically, from space itself. The comparison with Einstein's original idea of a 'cosmological constant', an irremovable repulsive force built into the texture of the universe, has been difficult to avoid, although for half a century it was common opinion that its addition to relativity theory was *ad hoc*. While little is certain about this accelerative force, one thing seems clear: As one physicist puts it, 'the energy density associated with the [new] cosmological constant is not possessed by matter or radiation, but by 'empty' space.'"[22]

DETERMINANT OF THE SPEED OF LIGHT

Modern science is divided over whether light is comprised of waves or corpuscles. Metaphysics looks at the thing differently, not from the phenomena it manifests, but from the perspective of being.

"For there are diverse degrees of entity according to which there correspond diverse manners of 'be' (*modi essendi*), and according to these degrees different things are classified."[23]

Metaphysics recognises ten special modes of being in two categories, *substance* and *accident,* as I have detailed in the opening chapter.

Is light a *substance* or an *accident*? Does it 'be in itself', or 'be in another'? Aristotle writes:

"[L]ight is the activity... of the transparent forasmuch as it is transparent... Light is, as it were, the proper colour of the transparent and exists whenever the... transparent is excited to actuality by the influence of fire, or something resembling 'the uppermost

---

22   Christopher A Decaen, *Aristotle's Aether and Contemporary Science,* op. cit., Pt III C

23   St Thomas Aquinas, *Disputed Questions about Truth* q. 1, a. 1: Sunt enim diversi gradus entitatis, secundum quos accipiuntur diversi modi essendi, et iuxta hos modos accipiuntur diversa rerum genera.

body'; for fire, too, contains something which is one and the same with the substance in question... [Yet] light is neither fire, nor any kind whatsoever of a body, nor something given off by any kind of body—for in such a case it would itself be a kind of body. It is the presence of fire, or something resembling fire, in what is transparent. It is certainly not a body, for two bodies cannot be present in the same place."[24]

St Thomas comments:

"But light (*lux*) differs from heat in this that light is a quality of first altering body which has no contrary: wherefore neither does light have a contrary, whereas heat does.[25] And because there is nothing contrary to light, it is not possible for there to be a contrary disposition in its recipient: and because of this its matter, the transparent body, is always immediately disposed to its form. That is why illumination occurs instantaneously, whereas what can become hot only becomes so by degrees. The participation or effect of light in a diaphanum is called "luminosity" (*lumen*)..."[26]

Both philosophers distinguish the light in a source (such as the Sun) from the light in the atmosphere. As can be seen above, St Thomas calls the former *lux* and the latter *lumen*.[27] He follows Aristotle in saying that *lumen* is to the diaphanous (*scil.* air) what colour is to a bodily surface (*corporis terminatum*). The colour in each is latent and activated by light from a source, *lux*. For colour to be visible—to act upon the organ of sight—the medium must have light (*lumen*) in it.[28]

Both philosophers deny that light travels through its medium like an arrow shot from a bow. They teach, rather, that light is the activation of a disposition present in the medium (whether transparent or

---

24   Aristotle, *De Anima*, Bk 2, Pt. 7

25   St Thomas is not saying that light has no contradictory; darkness is its contradictory. He is saying it has no *contrary* which is the opposition between two of the same genus opposed to each other, as red and green are contraries—the one, when adopted, drives out the other – which may occur, as he says, by degrees.

26   *In II De Anima*, lect. 14, nn. 6-7.

27   Though occasionally he used *lumen* when he might have used *lux*.

28   Which is not to deny that the source itself may be coloured, as the Sun is yellow-ish; as the star Sirius is blue-ish.

diaphanous). The modern scientist may discount their approach because he can demonstrate that light *does* progress, but they are right. There is no process, for instance, in the lighting of Earth's atmosphere: it is lit instantaneously. That light may require time to effect its qualitative activity over vast distances occurs only because the transparent and the diaphanous suffer from matter's inertia.[29] The fifth property follows, I contend, on this inertia. It answers, moreover, the question why the speed of light is not infinite but determinate—limited to 299,792.458 km/s *in vacuo*. It is *aether* that determines it.[30]

Aristotle did not distinguish the heavenly substance from the heavenly lights it contains—Sun, Moon and stars. He conceived of *aether*, then, as *the luminescent;* and St Thomas reasoned that it was in virtue of this quality of luminescence that *aether* acts upon ordinary matter. Modern science seems to show that in addition to *aether's* potency to activation by light, it is in potency to activation by various other forms of electromagnetic energy.

Every *substance* determines the *qualities* it bears. To illustrate, the (*accidental*) form of heat induced in water (to which form it is in potency) is determined by the nature of water.[31] So the *accidental* form of light induced in *aether* (to which it is in potency) is determined by the nature of that *substance*. The value *c* does not, as science thinks then, indicate the speed of light but the speed at which *aether* determines the development of that *quality*.[32] It is, likewise, the speed at which *aether*

29   I am mindful of the findings of Michelson and Morley. I am not using the term 'inertia' here as if of a common material being, what Einstein, cited in a later footnote means by the expression 'ponderable medium'. Unlike common matter *aether* has no weight. Yet because it is material (and not immaterial) it cannot operate unlimitedly, as, e.g., by permitting the propagation of light at infinite speed.

30   Science can demonstrate that the speed of light slows in different transparent media, as e.g., in water, in glass and in diamond. The alteration in speed is indicated by the refractive index of the medium. That of typical glass is 1.5. This means that the speed of light in glass is 1/1.5 = 0.67 times its speed in a vacuum. That of diamond is 2.41 giving a speed of only 0.415. If the argument advanced here avails, the atomic or molecular structure of each such medium serves to impede the facility of luminiference of the *aether* which permeates its structure. (Though note that conceiving of *aether* on analogy with some fluid misunderstands its nature.) For substances which admit of no transparency, the atomic structure must provide a complete impediment to this facility *qua* light. But not necessarily in respect of other forms of electromagnetic energy; e.g., x-rays.

31   And is not any heat at all, but the heat peculiar to water, bound by that substance's limitations.

32   The letter *c* that Einstein uses to stand for the 'speed of light' is taken from the Latin *celeritas* meaning 'speed', or 'swiftness'.

determines the development of the other *qualities* of electromagnetic energy to which it is in potency.

In his celebrated formula Einstein lays down that the equivalence between mass and energy is a function of c—c^2 to be precise. If c is the speed at which *aether* determines the development of the *qualities* represented by the various species of electromagnetic energy, and one or other of these is the means whereby *aether* binds the nucleus and associated electrons in every atom, the equivalence between mass and energy as a function of c makes sense. Einstein's formula is misleading in appointing c as a property of light. Take it as a property of the matrix in which all atomic structure subsists and is determined, *aether,* and its true significance appears.

In line with the thesis advanced here, the scientific expression *in vacuo* is to be countered with the metaphysical *in aethere*. The scientist means by his expression that all other matter is excluded: he asserts a void. This is impossible; where no other (common) matter is present, *aether* is. All generation and corruption, all material activity then is, following this argument, *in aethere*.

The Mode of Aether's Involvement

There is a problem for metaphysics—if not for science—the demands of the doctrine of *hylemorphism*.[33] Science looks at substantial change from the point of view of the phenomena detected. The formation of water occurs by the combination of what it identifies as one atom of oxygen with two of hydrogen. Metaphysics looks at the business from the perspective of being. Water is formed when the *substantial form* of water combines with *prime matter*. It allows that the *form* of each of two *substances* (e.g., oxygen and hydrogen) may be corrupted in favour of the *form* of another (water) in the substantial change but the *matter* stays the same, which truth science demonstrates by comparing the weight of the materials before and after the substantial change. The old substances are replaced by the new. For metaphysics *substantial form* is the determinant of the nature of the new substance and on this the nine accidents follow, the first of which, *quantity*, "quantifies and... materialises its subject by extending it and ordering its material parts".

---

33   A word derived from Greek signifying the compound of matter and form.

No other cause determines the nature of the new substance. How, then, could *aether* be involved in the process?

Metaphysics allows the involvement of a cause *per accidens*, a cause which contributes to the effect by removing something prohibiting the *per se* cause from producing its effect—a *conditio sine qua non*.[34] Thus, the sea is not a *per se* cause either of the becoming or of the subsistence of the fish within it: yet it is an essential condition of both. In the same way, I suggest, *aether* is an essential condition of the subsistence—and in the case of living things, the generation—of all common material beings. Similarly, the opening of a tap is not a *per se* cause of the water flowing through it, the efficient cause is gravitational force, but the opening of the tap is an essential condition. The process called *catalysis* where the presence of some element or compound facilitates a chemical reaction operates in an analogous fashion.[35] *Aether* may be understood as contributing to the existence of a common material *substance* by cooperating with its first accident, *quantity*, in facilitating the ordering of its material parts at the molecular level. It does so extrinsically and instrumentally, an efficient cause.

Aether and Time

Time is the number, or measure, of movement or change. While *aether* is clearly moveable *per accidens*—for it adapts to the movement of every element of common material being—it is *per se* immovable; immutable, incapable of change; and therefore outside time. This superiority has, I suggest, a consequence. Since first he walked the Earth man has measured time according to the rotation of the Earth around its axis, of the earth around the Sun, and of the Moon around the Earth.[36] The most perfect method of keeping time now is by means of atomic resonance.[37] If *aether* governs not only the movements of the heavenly bodies but

34   In V *Metaphysics*, lesson iii

35   Yet a catalyst is not essential to the reaction. Science now identifies different types of catalytic operation including instances in which the catalyst is involved in the reaction and partly, or wholly, consumed. The original conception of *catalysis* is, however, of a material substance that accelerates a chemical reaction without being consumed in the process.

36   Or in the regularity of repetition of the same sequence in consequence of these rotations.

37   Cf. http://en.wikipedia.org/wiki/Atomic_clock

those of the atoms that make up every element of common material being then it is indelibly involved in the reckoning of time.

* * *

There is one final, metaphysical, comment to be made about this extraordinary substance. St Thomas says:

> "The celestial bodies are far from us not only according to quantity of spatial distance, but even more so in that few of their accidents fall under our senses, while it is nevertheless connatural to us that we proceed from accidents, i.e., sensibles, to knowing the nature of some thing... But the accidents of the celestial bodies are of a different notion altogether [alterius rationis] and are wholly disproportionate to the accidents of inferior bodies." [38]

In particular, because *aether* is ungenerable and incorruptible, the accident *quality* does not manifest in it the characteristic change of properties it works in common material being.[39] This explains why in *aether* light does not light it, nor does heat heat it.

Of the other seven accidents, *when* and *where*—that is, time and place—cannot be attributed to it. *Relation* may be said of *aether*, but only analogously. Its relation to all creation is like the sea to the fish that subsist in it, as container to the contained; as the essential condition to material existence. Of the remaining four accidents, *action* may be attributed to it, but not *passion*, because *aether* acts yet is not acted upon. But neither *situs* nor *habitus* are applicable; *situs*, because it consists in the order of the parts of its subject, but since the parts of *aether* are not detectable, neither is their order; and *habitus*, because this is taken from something outside the subject (yet not a measure of it); but nothing is extrinsic to *aether*. Rather, *aether* is extrinsic to everything else.

---

38   *In II De Caelo*, l. 4. n. 3

39   *De Caelo* I, 3. Aristotle says: "qualitative states and dispositions do not come into being without changes of properties. But we see that all natural bodies which change their properties are subject without exception to increase and diminution."

The Michelson-Morley Experiment was a Success

The Michelson-Morley experiment showed that the speed of light was constant in all frames of reference. In 1905 Albert Einstein published his special theory of relativity which drew on this conclusion. Einstein announced shortly after that a luminiferous *aether* was outdated. Fifteen years later, however, he recanted.

> "More careful reflection teaches us... that the special theory of relativity does not compel us to deny ether... [W]e may say that according to the general theory of relativity space is endowed with physical qualities; in this sense, therefore, there exists an ether... According to the general theory of relativity, space without ether is unthinkable; for in such space there not only would be no propagation of light, but also no possibility of existence for standards of space and time (measuring-rods and clocks), nor therefore any space-time intervals in the physical sense. But this ether may not be thought of as endowed with the quality characteristic of ponderable media, as consisting of parts which may be tracked through time. The idea of motion may not be applied to it."[40]

This view has remarkable resonance with the teaching of Aristotle and St Thomas.

Christopher A Decaen closes his paper with what he calls "the resuscitation of *aether*" by contemporary science. In the working out (by Einstein's successors) of the General Theory of Relativity and in the field of Quantum Electrodynamic Theory—that is, in both the *macro* and the *micro* areas of its concerns—science is moving back to the view that some form of *aether* is essential.

What do we conclude? Far from being a failure, the Michelson-Morley experiment was a success. It established at the scientific level what Aristotle and St Thomas had maintained at the philosophic, namely, that *aether* does not share the accidents of common material being, and is immutable. It is reasonable to argue that it is this immutability that grounds the immutability of the speed of its proper accident, light.

---

40   In a lecture meant for his inauguration at the University of Leiden in 1920. Quoted in Albert Einstein, *Sidelights on Relativity*, trans. G B Jeffrey and W Perrett, New York, (Dover) 1983, 13, 15. And cf. footnote 106 in Christopher A Decaen, *Aristotle's Aether and Contemporary Science*, op. cit.

There would seem to be one other corollary. The experiment showed the falsity of the thesis that matter is 'nearly all free space' and that if that 'space' was removed the Earth 'would be reduced to the size of an orange'. The 'space' is not removable (save in the scientific imagination); what the scientists call 'space' is *aether,* the essential condition of its atomic structure, without which no element of common material being could exist.

One may wonder whether 'dark matter', 'dark energy', 'black holes' and 'curved space' posited by scientists represent realities or whether they are merely conceptual constructions of scientists to account for phenomena not otherwise explicable. One may wonder, too, whether science would continue to posit them if it adopted Aristotle's understanding of the heavenly body.

One thing is certain; scientists will make exponential advances in understanding the majestic world of creation if only they will rid themselves of materialism's stifling mindset.

* * *

One critic has offered a studied objection to the above thesis as follows:

> "If *aether* is one of the media that transmits light as does air, water, etc., then we could just as well say that any one of them is the one cause or explanation. There is no reason to privilege *aether.* In any case *aether,* as Aristotle conceives it, is a body (no doubt very subtle—aethereal) not a pure quality. I fear that you are hypostasising the qualitative potential of the medium which Aristotle says is common to air, water, *aether* etc. We may not have a name for [this potency] but that does not mean we have to identify it with one of these corporeal bodies which function as media of light, heat etc. The heat in water and air, for instance, is not due to one so that I need to say that the heat in water is because there is air in it. The form of heat or light in the different corporeal media is not the same but similar only. The heat of the air is not the same heat as that of the water though they are necessarily from the same source, and a likeness of it."

By "hypostasising" the critic means "treating as a substance what is in truth an accident".

The objection is a good one. I think it is to be answered in this way. Neither air nor water nor glass nor any other diaphanous medium can be the means of transmission of light from the Sun. But *aether* is. Hence *aether* is, to use the suggested term, "privileged". *Aether* is a material substance though not in the sense in which we understand that term of common material being. It can only be called a body analogically, somewhat after the fashion in which we speak of the sea as a body of water. The qualitative potential in *aether* cannot be the same as that in air, water, or glass, etc. because *aether* is not a material substance in the way those things are material substances, but analogically.

Thus, in contrast with the manner in which these qualities appear in air, *aether* is not lit by the light, nor heated by the infra-red radiation it carries. Aristotle lays the ground for establishing why this is so in *De Caelo* I, 3:

> "Alteration is movement in respect of quality... [Q]ualitative states and dispositions do not come into being without changes of properties. But we see that all natural bodies which change their properties are subject without exception to increase and diminution."

But *aether* is immune from generation or corruption and is incapable, therefore, of increase or diminution. Hence it is not affected by the qualities it carries.

In a memorable passage in his theological treatise, *The Ascent of Mt Carmel,* St John of the Cross illustrates the truth that light is invisible *per se.*

> "If a ray of sunlight should be entirely cleansed and purified of all dust particles, even the most minute, it would appear totally obscure and incomprehensible to the eye, since visible things, the object of the sense of sight, would be absent. Thus the eye would find no images on which to rest, because light is not the proper object of sight, but only the means through which visible things are seen. If there is nothing visible off which the ray of light can reflect, nothing will be seen. If the ray, then, were to enter through

one window and go out another without striking any quantitative object, it would be invisible."[41]

Light and heat are conveyed to us from the Sun through *aether* invisibly as, science tells us, electromagnetic energy of different wavelengths. Heat is conveyed primarily as infra-red radiation which affects all the material bodies with which it comes in contact. The heat so conveyed is passed in these bodies by conduction or convection.[42] Each form of electromagnetic energy is conveyed at *c*, 'the speed of light', and continues to be conveyed at that rate in Earth's atmosphere and in other diaphanous media. The speed of its conveyance is qualified, however, by the atomic structure of the medium through which it passes, the extent of the suppression indicated by the refractive index of the medium in question. Science takes as a standard for refractive index the speed of light's development *in vacuo*. For practical purposes, however, it uses air at a standard temperature and pressure.[43] Air too has a refractive index relative to that of a 'vacuum' which seems to be of the order of 1.0003, marking a fractional slowing of the speed of light in Earth's atmosphere. This, doubtless, holds true also for all other species of electromagnetic energy.

We take *aether's* immensity for granted. The light from *proxima centauri* (a red dwarf part of *alpha centauri*, the closest star group; there are three stars so closely aligned that the human eye cannot distinguish them) travels some 40 million million [299,792 x 31,536,000 x 4.22] kilometres to reach our eyes, of which only the last twenty or so are constituted by Earth's atmosphere. In the whole universe no element of common material being is more extensive than *aether*. But this understates the reality. The universe and *aether* are convertible: *aether is* the universe!

Christopher A Decaen has this to say in his paper:

"If *aether* is incorruptible two conclusions follow right away, one pertaining to its substantial principles and the other pertaining to

---

41   Book II, ch. 14, n. 9. Translation by Kieran Kavanaugh O.C.D. and Otilio Rodriguez O.C.D., *Collected Works of St John of the Cross*, Washington D.C., 1979, p. 145.

42   The Principle of Reception applies—*Quidquid recipitur, per modum recipientis recipitur*. Whatever is received, is received according to the mode of the recipient.

43   Cf. http:/en.wikipedia.org/wiki/Refractive_index

its qualities. First, *aether's* prime matter and substantial form must be so perfectly united that the latter must actualize and thereby exhaust the potency of the former, insofar as an incorruptible body by definition must lack the potential to become anything else; *aether* must possess a 'certain total and universal perfection' that thoroughly fulfils its potency for existence. Indeed, if one were not to distinguish fulfilled and unfulfilled potencies, one might be tempted to say that the heavenly substance *has no* prime matter. More accurately, however, one should conclude that, unlike sublunary composites, *aether's* prime matter is always perfectly fulfilled, so it is inseparable from its form, and in this sense is not really distinct from it. Likewise, since its prime matter would not be a principle of *aether's* coming to be, but only of its being, it would not be the same sort of prime matter that is a principle of mundane substances (which is a principle both of coming to be and of being); it would be called prime matter only analogously."[44]

My early thinking was that *aether* was only an instrument of light's transmission—that is, a pure instrument. I accepted, however, in accordance with the mind of Aristotle and St Thomas, that light is a quality and *aether* is in potency to that quality. It is almost as if as regards light and the other species of electromagnetic energy *aether* is not material. I think then, reasonably, that one might adopt the criticism offered and speak of *aether* almost as if it was an hypostasised accident.

Aether is, in my view, the matrix of all physical reality though by 'matrix' here I do not mean that it is the source of being of common matter after the fashion of a mother liquor towards the crystals that grow in it. A fish cannot exist except in water which is a *per se* cause neither of its coming into existence (becoming) nor its existence (being). Yet water is an essential condition, a cause *per accidens,* in respect of both. In the same way, I contend, no material thing comes into existence, or subsists, but in *aether* which is as essential to it as is water to the fish.

---

44 *Aristotle's Aether and Contemporary Science,* op. cit. Decaen quotes a phrase from St Thomas's commentary on Aristotle's De Caelo (*I De Caelo* VI, 6): "huic autem materiae vel subiecto non est nata inesse alia forma, sed forma sua replet totam potentialitatem materiae, cum sit quaedam totalis et universalis perfectio."

What follows? *Aether* must be ontologically prior (i.e., prior in the order of reality) to all common material being.

3. LIGHT

In the beginning God created the heaven and the earth. And
the earth was void and empty and there was darkness over the
face of the deep. And the spirit of God moved over the waters.
And God said: Let there be light. And light was made.

Genesis I: 1-3

WHAT REALITY IS more critical in our lives than light? It is hardly less important than existence itself. Yet how shall we understand it? Or define it? Its reality is only balanced by its intangibility. Science can hypothesise about light, but its observations are limited to, well... observations. It cannot tell us anything about light's essence. What has metaphysics to say?

First it should be remarked that modern experimental science and metaphysics are not mutually incompatible. Each discipline has information and analysis it can provide the other to their mutual advantage and to the benefit of mankind. But science suffers from an intellectual problem caused by its adherence to an erroneous philosophy. By definition, nothing is *no thing*: it does not exist. But modern science allows that non-being does exist, for by the names 'space', 'vacuum', or 'void', it conceives of non-existence as if it is a reality.

The scientist is also insouciant about causes. The root of his conduct lies in another intellectual defect, one that tends to balance materialism's facile postulates, *subjectivism*. Subjectivism breaches a fundamental rule of logic. It treats being which exists only in mind, mental being, as *real* being and inclines its adherents to move between the one and the other—between the objective and the subjective orders—as if they were interchangeable. Thus the modern scientist thinks nothing of interpolating in the intellectual consideration of critical questions, the operations of his imagination. Because he can *imagine* nothing existing, he is prepared to allow that it does exist. That he is admitting the impossible does not trouble him.

There is no such thing as a subsistent nothing. Whatever there is between the Earth and the Sun, between Earth and the furthest star, between the atom and its electrons, even if experimentally undetectable, it must be *something*: a material substance.

Both Aristotle and St Thomas teach that light is an accident, a *quality*. Just as we never experience heat save in something hot, neither do we experience light save in something lit. We perceive light in diaphanous things; in Earth's atmosphere and in water; in fluids like turpentine or alcohol; in glass, diamond, and other precious and semi-precious stones. Since each appears to participate to some extent in light's nature, it is tempting to insist that light is a quality of each.[45] A closer consideration of diaphanous substances reveals that they resemble the opaque in that they manifest colour, but much more subtly than the opaque. They may be reduced, then, to the particulated (or rarefied) opaque. Accordingly, though we speak as if light is a quality of the diaphanous, strictly it is not, even of the most etiolated. Of what substance, then, is light properly the quality?

As mentioned in Chapter 2, St Thomas distinguishes light into two categories; the light emitted from a source (which he calls *lux*); and reflected or dissipated light (which he calls *lumen*). Now, all *lumen* is reduced to *lux* as every effect is reduced to its cause. Of what substance, then, is *lux* the quality? St Thomas teaches:

> "Light is a quality of first altering body, the most perfect and least material of all bodies…"[46]

This is Aristotle's 'first body', or 'heavenly body', or *aether*, whose attributes we have set out above.

Aristotle says:

> "Clearly there exists something transparent… But transparency does not depend on air or water as such, but on the same quality

---

45 In contrast, the opaque surfaces of material things do not appear to participate in light's nature save in rare instances, luminescent bodies whose luminescence is stimulated by the action of light, or by what science refers to generally as "electromagnetic energy". Rather, light has the effect of manifesting in them their proper quality, colour. This may be thought a participation in light's nature but it is so only *secundum quid*, i.e., in a secondary fashion.

46 *In II De Anima* lect. XIV, n. 24.

being found in both, and in the eternal sphere above as well. Light is the act of [the transparent]... [It] is a kind of colour of the transparent in so far as this is actualised by fire or something similar to the heavenly body; which contains indeed something of one and the same nature as fire.[47]

St Thomas comments:

"[J]ust as the corporeal elements have certain active qualities through which they act, so light is the active quality of the heavenly body through which it acts. It falls within the third species of quality, like heat."[48]

Light is then, according to these philosophers, a quality not of any earthly substance but of 'the heavenly body', Aristotle's *aether*.

The thoughtful reader will immediately see that there is a problem. If *aether* is confined to the heavens, how can light, *aether's* proper quality, be seen in Earth's atmosphere? There appear to be two explanations. Either diaphanous substances (air, water, glass, precious stones, etc.) participate in the quality proper to *aether*, or else *aether* is present in the midst of diaphanous substances. In other words *aether* is not, as the philosophers thought, confined to the heavens but is universal.

The First View

Aristotle expresses the first view where, in the passage quoted above, he says:

"Neither air nor water is transparent because it is air or water. Each is transparent because there is contained in it a certain quality which is the same in both and is also found in the eternal upper body."[49]

---

47   *De Anima*, Bk. II, Pt. 7

48   In *II De Anima* lect. XIV, n. 22: Unde dicimus, quod sicut corpora elementaria habent qualitates activas, per quas agunt, ita lux est qualitas activa corporis caelestis, per quam agit, et est in tertia specie qualitatis sicut et calor.

49   *De Anima*, Bk. II, Pt. 7.

Commenting on this, and on earlier material where Aristotle deals with colour and the visible, St Thomas says:

> "[I]t is evident that neither air nor water nor anything of that sort is actually transparent (*transparens*) unless it is illuminated. Of itself the transparent (*diaphanum*) is in potency to both light and darkness (the latter being a privation of light) as primary matter is in potency both to form, and the privation of form."[50]

Where Aristotle uses the one Greek word *diaphanes* to signify 'the transparent', St Thomas introduces a second in Latin: he distinguishes *transparens* 'the transparent', from *diaphanum* 'the diaphanous'.[51] He does so, it would seem, to contrast the heavenly matter, *aether,* with the mundane (air, water, glass, etc.). His distinction of the heavenly matter from the earthly may be seen in this extract from his commentary on the second book of Aristotle's *De Caelo* quoted above, which I repeat here for the sake of convenience.

> "The celestial bodies are far from us not only according to quantity of spatial distance, but even more so in that few of their accidents fall under our senses, while it is nevertheless connatural to us that we proceed from accidents, i.e., sensibles, to knowing the nature of some thing... But the accidents of the celestial bodies are of a different notion altogether [*alterius rationis*] and are wholly disproportionate to the accidents of inferior bodies."[52]

St Thomas takes a further step as said above: he distinguishes the light produced in a source such as the Sun, or stars, or earthly fire, which he calls *lux*, from the light which is found in diaphanous substances like air or water which he labels *lumen*. He ascribes *lux* chiefly to the

---

50 *In II De Anima*, lect. XIV, n. 7

51 Though, in places, he seems to see no difference between the two, as e.g., in Lecture XIV, n. 5: "The diaphanous is the same as the transparent (e.g. air or water)..."

52 *In II De Caelo*, lect. IV, n. 3. St Thomas uses the expression "the celestial body" in two senses here as is explained hereafter.

transparent and restricts *lumen* to the diaphanous.[53] These distinctions are a valuable addition to Aristotle's thinking, and as we will see, they seem to be confirmed by the discoveries of modern science.

The Second View

The second view (the present author's thesis) grounded in the assertion of *aether's* universality is that it is *aether*, and only *aether*, which is the medium of light's transmission. Its universality is hinted at by Christopher A Decaen in his paper *Aristotle's Aether and Contemporary Science* where he considers modern science's postulate of a 'vacuum' as part of the theory of quantum electrodynamics. I contend that even in the diaphanous (air, water, glass, diamond and other stones) where light is dispersed (i.e., *lux* becomes *lumen*) and colour appears, it is *aether* present in the diaphanous media which is the vehicle of light's transmission and which enables us to see. Pursuant to this thesis, light, that most immaterial of material qualities, is proportioned (and proportioned only) to the most immaterial of material substances.

The chief objection to the thesis is this: it is impossible for two bodies to be in the same place at the same time. This principle does not, of course, prevent the contiguous. It allows for fish moving through the sea, for silt muddying the waters of a river, for birds flying through the air, and dust penetrating the Earth's atmosphere. Why, then, should it prevent the contiguity with every element of common material being of a substance so subtle it cannot be detected experimentally? Moreover, if material bodies are, according to the discoveries of science, 'mostly empty space' and this 'empty space' is *aether*, the objection does not avail.

---

53   *In II De Caelo*, lect. IV, n. 3. An accident of history may have served to precipitate the distinction. St Thomas did not see Aristotle's original Greek text, but a Latin translation effected by his fellow Dominican, William of Moerbecke. It may have been William exercising the prerogative of the translator who first rendered Aristotle's one term in the Greek with the two terms used by St Thomas in the Latin. There is evidence of this in those places where St Thomas quotes Aristotle apparently from the Latin text before him, as e.g., at L. XIV, n. 6, "Secundo determinat de lumine, quod est actus eius, ibi, *lumen autem est huius actus et cetera*"; and at L. XIV, n. 7: "Deinde cum dicit *lumen autem...*" I have placed what St Thomas attributes to Aristotle in italics.

* * *

It seemed to St Thomas, as it had seemed to Aristotle 1,600 years prior, that 'the heavenly body' was comprised not only of the transparent matrix but also of the Sun, stars and planets ('the wandering stars') it contained. Thus, in Lecture XIV n. 6 of his Commentary on Aristotle's *De Caelo*, he says:

> "For at least some celestial bodies are manifestly transparent. We should not be able to see the fixed stars of the eighth sphere unless the lower spheres of the planets were transparent or diaphanous."[54]

And, in Lecture XIV n. 7 he says:

> "[T]o be enlightened and illuminating is common to fire and the celestial body, just as to be diaphanous is common to air and water and to the celestial body."[55]

He uses 'celestial body' (*corpus caeleste*) in two different ways here. The first refers to lucent bodies (*corpora lucentia*) such as Sun, stars and planets. The second refers to the heavenly matrix in which they appear. Here he uses the singular in each case. In the earlier passage from the same text cited above (lect. IV, n. 3), he uses the plural for the first and the singular for the second.[56]

The reasoning that the heavenly substance was not simply (i.e., un-mixedly) transparent seems to be this, that it contained the heavenly lights. But we, blessed with the advantages of modern science, know

---

54   *In II De Caelo*, lect. XIV, n. 6. Manifestum est enim aliqua caelestia corpora esse diaphana. Non enim possemus videre stellas fixas, quae sunt in octava sphaera, nisi inferiores sphaerae planetarum essent transparentes, vel diaphanae.

55   *In II De Caelo*, lect. XIV, n. 7. Esse enim lucens actu et illuminativum, commune est igni et corpori caelesti, sicut esse diaphanum est commune aeri et aquae, et corpori caelesti.

56   *Corpora* autem *caelestia* non ita sunt longe a nobis tanto... Hanc autem elongationem dicit multo maiorem esse quam localem: quia si consideremus localem distantiam, aliqua proportio est distantiae qua distat a nobis *corpus caeleste*, ad distantiam qua distat a nobis aliquod inferiorum corporum, puta lapis aut lignum, et utraque distantia est unius generis; sed accidentia *caelestium corporum* sunt alterius rationis, et omnino improportionata accidentibus inferiorum corporum. [emphases added]

two things (among innumerable others) that the two philosophers did not know, namely:

- that 'the heavenly bodies', St Thomas's *corpora lucentia,* are no different in composition from those on Earth and, in the case of the Sun and stars, they differ from mundane fire only in degree of heat (and source) and mass; and,

- that the distances of the stars (and even of the planets and the Sun) from our eyes are inconceivably great.

What follows? First, 'the heavenly bodies' are not part of the heavenly substance at all, but are located in that substance like the rest of common material being. Second, it is impossible that the lights of 'the heavenly bodies' could be conveyed to our eyes if the substance through and by which they are transmitted to us shared in the nature of even the most refined of diaphanous substances, air. For even at its most refined, the diaphanous will eventually obscure the light that passes through it. *Aether* must, then, as Christopher A Decaen has argued, be transparent by essence.

Diaphanous substances are not *per se* transparent. Indeed, they seem to be a mixture of the transparent and the opaque. The air in Earth's atmosphere when lit by the Sun always displays colour--now blue, now gray, now pink or red—depending on the atmospheric conditions obtaining, the colour of the sea (itself determined by the presence or absence of cloud cover) and the admixture of moisture, dust or other matter. In other words, diaphanous substances are only qualifiedly transparent. *Aether*, on the other hand, is transparent by essence. It is not lit by the light it carries.[57] Indeed, light is invisible until it strikes a *corporis terminatum*, i.e. the surface of an extended common material body, whether opaque or diaphanous. This characteristic invisibility of light is implicit in the teachings of Aristotle and St Thomas but, so far as the writer is aware, is never expressed by either of them.

---

57   Further consideration will show that *aether* is not affected in any way by the other species of electromagnetic energy it carries. Nor, indeed, are any of these elements of energy detectable in it. They are detectable only, after transmission from their source, in a recipient.

Here is a synopsis of the teachings of the two philosophers and of the discoveries of modern science together with my interpretation of their significance.

- Light is not a *substance*; i.e., it is not, as one scientific hypothesis has it, a body.

- Light is an accident, a *quality*, not of any substance at all, but of *aether*, and only of *aether*.

- Light in a source is properly termed *lux*.

- In *aether* (the transparent by essence) light is invisible.

- Light becomes visible only when it strikes a *corporis terminatum*, the surface of some element of common material being, whether opaque or diaphanous, or its proper organ of reception, the eye.

- Light in the diaphanous may properly be termed *lumen*.

- Light is to the diaphanous what colour is to the rest of common material being.[58]

- Light is an active quality (in particular, for living things), as *aether*, its proper substance, is an active substance.[59]

- Light (i.e., white light) contains all colour virtually.

- Light activates colour in a *corporis terminatum* via the instrumentality of the diaphanous which acts as a virtual prism.[60]

---

58   Here I modify the teaching of St Thomas that contrasts light *in the transparent* with colour in common material being by insisting more radically on his distinction between *the transparent* and *the diaphanous*. Light (i.e., as *lux*) is invisible in *aether*, the transparent; it is visible, as *lumen*, in the diaphanous (air or water). It is to its presence *in the diaphanous*, I argue, that St Thomas was referring.

59   For a consideration of the activity of the heavenly substance, see the author's *Science and Aristotle's Aether*, at http://www.superflumina.org/PDF_files/aether_science.pdf

60   Underwater photography under artificial light at depths in the ocean which prevent the penetration of atmospheric light shows that other diaphanous materials (e.g., sea water) provide this function in the same way as air.

Is Light In Fact Invisible?

It is possible that, like all the other forms of what science calls 'electro-magnetic energy' transmitted in *aether*, light is invisible *per se*. If this be the case, what we call 'light' (*lux*) is simply primary colour, white, in varying degrees of splendour excited in a source by this invisible active quality. It is dissipated (as *lumen*) in the diaphanous which acts as a virtual prism eliciting the colour present in every bodily surface (*corporis terminatum*). When *lux* strikes a *corporis terminatum* in the absence of the diaphanous it is reflected only in primary colour (white) though the surface's proper colours affect the reflection by diminishing its purity so that the object lit appears, as we would say, in black and white. This phenomenon is most manifest in photographs of the nuclei of comets taken from unmanned space craft.[61] It is less so in photographs of the Moon's surface and of the exterior of artificial 'space stations' where there exists a residual atmosphere.

St Thomas considered this ancillary issue in his Commentary on the *De Anima*:

> "[S]ome have simply identified light with the manifestation of colour. But this appears clearly to be false in the case of things that shine by night while their colour is hidden."[62]

But this objection will not stand if white is understood as primary colour, that is, the matrix in which all colours are virtually contained. Before the time of Newton it was thought that a prism produced colour from white light. By using prisms against each other Newton demonstrated that all colours emanate from, and are virtually contained in, white light. In this he demonstrated experimentally the truth enunciated by St Thomas 450 years before, that:

> "light (*lux*) is, in a certain manner, the very substance of colour..."[63]

---

61 See e.g., photos on pages 18, 25, 65 & 138 of David J Eicher, *Comets! Visitors from Deep Space*, CUP, New York, 2013.

62 *In II De Anima* Bk II, lecture XIV, n. 21 § 419. Quidam vero dixerunt quod lumen non est nisi evidentia coloris. Sed hoc aperte apparet esse falsum in his quae lucent de nocte, et tamen eorum color occultatur.

63 *In II De Anima*, Lecture XIV, n. 28

Should this claim that light is *per se* invisible be demonstrated to be true, so close is the identification between light and its proper effects there could be little difficulty in continuing to ascribe to these effects the name 'light'.

* * *

Confirmation in Divine Revelation?

The words quoted in the epigraph to this chapter are the very first words in the Book of *Genesis*, the first book of Almighty God's revelation to mankind. The popular rendition of the passage speaks of 'the heavens' but St Jerome used the singular 'heaven' in his translation from the Hebrew in the Latin Vulgate. The popular interpretation has it that 'the heavens' refers to the great and lesser lights of the night sky and 'the earth' refers to the planet on which we live. (A critic might object that the interpretation places the light-producing heavenly bodies *before* the creation of light, though the answer may be that the heavenly bodies were created before their Author enabled them to produce light—and certainly they are prior, ontologically, to the light they produce since do follows be.)

But there is another interpretation consistent with the arguments advanced here. The words were written to a people of comparatively limited knowledge, one lacking any conception of the scientific realities which would come to be known. As Pius XII said of the first eleven chapters of the Book of *Genesis*:

> "[I]n simple and metaphorical language adapted to the mentality of a people but little cultured, they both state the principal truths which are fundamental for our salvation, and give a popular description of the origin of the human race and the chosen people".[64]

64   *Humani Generis* (August 12th, 1950) n. 38

Why, then, should not 'the heaven' properly be regarded as referring to the matrix, the great sea, in which the whole of material creation is established, Aristotle's heavenly body, or *aether*? A people but little cultured, lacking the sophistication of knowledge that has abounded since the late 19th century, could not but regard 'the earth' as referring only to the lands and seas with which their existence was circumscribed. But the emergence of that knowledge may justify the extension of the phrase 'the earth' to the whole of common material being, the stars in their billions—including our Sun,—all the planets, not just our Earth, all the moons, asteroids, comets and other entities with which the Almighty has peopled the universe.

The order indicated in the text is significant: first God created 'the heaven', then 'the earth': first He created the setting, then all the entities that inhabit the universe. If this is the case *aether* is ontologically prior to all common material being.

* * *

APPENDIX

LIGHT: ARISTOTLE & ST THOMAS

Set out below is the teaching of Aristotle in his treatise on the soul, *De Anima*, Book II, Chapter 7 (on *Sight and Its Object* and *How Colour is Seen)*, followed by St Thomas Aquinas's *Commentary* on the text (*In II De Anima*, Book II, lecture XIV and part of lecture XV). I have added some notes of my own to St Thomas's *Commentary* in an endeavour to reconcile with it the discoveries of modern empirical science.

One of the difficulties of the modern metaphysician is to place himself in the cosmological position of Aristotle and St Thomas. We take for granted so many discoveries about the Earth and the universe that we have difficulty in reducing our perceptions to the limitations of their knowledge of the natural world. There is, moreover, an inclination to reject their views because they lacked our scientific advantages. But from the little available they gathered much more about reality than our modern thinkers have been able to do because they were not concentrating on the appearances of things but on their essences.

The views expressed here are offered as a contribution to the natural philosophy of Aristotle and St Thomas for acceptance, amendment, or correction by better minds.

* * *

Aristotle's text [*De Anima* Book II, Chapter 7][65]

That of which there is sight is the visible; and the visible is colour, and also something which, though it has no name, we can state descriptively. It will be evident what we mean when we have gone further into the matter.

For the visible is colour, and it is this of which visibility is predicated essentially; not however, by definition, but because it has in itself the cause of being visible. For every colour is a motivating force upon the actually transparent: this is its very nature. Hence nothing is visible without light; but by light each and every colour can be seen. Wherefore, we must first decide what light is.

There is clearly something transparent. By transparent I mean that which is indeed visible yet not of itself, or absolutely, but by virtue of concomitant colour. Air and water and many solids are such. But transparency does not depend on either air or water as such, but on the same quality being found in both, and in the eternal sphere above as well.

Light is the act of this transparency, as such: but in potency this [transparency] is also darkness. Now, light is a kind of colour of the transparent, in so far as this is actualized by fire or something similar to the celestial body; which contains indeed something of one and the same nature as fire.

---

65   The translation is taken from the text in English reproduced in Kenelm Foster and Silvester Humphries, *Commentary on Aristotle's De Anima*, (Dumb Ox Books, Notre Dame, Indiana, 1994), a revised edition of a Yale University Press publication of 1951.

We have then indicated what the transparent is, and what light is; that light is not fire or any bodily thing, nor any emanation from a body—[if it were this last,] it would be a sort of body, and so be fire or the presence of something similar in the transparent.

For it is impossible for two bodies to exist in the same place at the same time.

Light seems to be the contrary of darkness; and the latter is the privation of this quality in the transparent. So it is plain that the presence of this is light.

Empedocles (or anyone else who may have said the same) was wrong when he said that light was borne along and extended between the Earth and its envelope, unperceived by us. This is in contradiction alike to sound reasoning and to appearances. Such a thing might happen unobserved over a small space: but that it should remain unnoticed from the east to the west is a very extravagant postulate.

Now that only can receive colour which has none, as only that which is soundless can receive sound. What is without colour is the transparent and the invisible, or what is barely seen, being dark. The transparent is precisely of this nature when it is not in act, but in potency. For the same substance is sometimes dark, sometimes light.

Not all visible things, however, are visible in light, but only the colour proper to each. There are certain things which are, indeed, not seen in light, but which produce a sensation in darkness, such as those which burn or are luminous. These are not called by any one term. Such are the fungi of certain trees, horn, fish-heads, scales and eyes. But the colour proper to each of these is not perceived. Why these things are thus seen is matter for another enquiry.

At present what is clear is that what is seen in light is colour; [and that] therefore it is not seen without light. For to be colour is to be able to move the transparent into act; and this act of the transparent is light. A plain proof whereof is that if one places on the sight itself a coloured object, it is not seen. But colour moves the transparent medium (e.g., air); and the sensitive organ is moved by this extended continuum.

Democritus put forward the erroneous opinion that if the medium were a vacuum, perception would be everywhere exact, even of an ant in the sky. This is, however, impossible; for only when the sensitive faculty is affected does vision occur. This cannot, however, be effected by the colour seen in itself. It must therefore be due to the medium. If there were a vacuum, a thing, so far from being perceived clearly, would not be seen at all. We have stated then, why it is necessary that colour be seen in light.

But fire is seen in both darkness and light: necessarily, for the transparent is made light by it.

* * *

St Thomas's Commentary [In II *De Anima* Lectures XIV, XV][66]

[Note: 'n...' indicates the paragraph reference to the text in the original Latin: '§...' indicates the reference in the Pirotta edition of St Thomas's texts. My annotations are sidelined.]

LECTURE XIV

n. 1 §399 Having distinguished the proper sense-objects from the common, and from those that are sensible incidentally, the Philosopher now treats of the proper object of each sense: first of the proper object of sight...

As to sight, he discusses, first, its object, and then, at 'At present what is clear', how this object comes to be seen. Touching the object of sight, he does two things. First, he determines what is the visible, dividing it into two. Secondly, at 'For the visible is colour', he deals with each. He says then, first, that, the proper object of a sense being that which the sense perceives *of itself exclusively*, the object of sense of which the special recipient is sight is the visible. Now in the visible two things are included. For while colour is visible, there is also something else which can be described in speech, but has no proper name. This relates to those things which can be seen by night such as glow-worms, certain fungi on oak-trees and the like, concerning which the course of this treatise will inform us more clearly as we gain a deeper understanding of the visible. But we must start with colour which is the more obvious visible.

---

66   The translation is based on that of Kenelm Foster and Silvester Humphries in their *Commentary on Aristotle's De Anima* cited above with amendments to expression in certain passages by the author.

n. 2 §400. Then, at 'For the visible', he begins to define both objects of sight, first colour and then, at 'Not all visible things', that of which he says that it has no proper name. As to colour he does two things: first, he shows what colour has to do with visibility; secondly, at 'There is, accordingly, something transparent' he settles what is required for colour to be seen.

First of all, then, he says that, colour being visible, it is visible of itself (*secundum se*), for colour as such is *per se* [essentially] visible.

n. 3 §401. *Per se* is said in two ways. In one way, when the predicate of a proposition falls within the definition of the subject, e.g., 'man is an animal'; for animal enters into the definition of man. And since that which falls within the definition of anything is in some way the cause of it, in such a case the predicate is said to be the cause of the subject. In the other way, on the contrary, it is said when the subject of the proposition falls within the definition of the predicate, as when it is said that a nose is snub, or that a number is even—for snubness is nothing but a quality of a nose; and evenness [is nothing but a quality] of a number which can be halved—and in these cases the subject is a cause of the predicate.

n. 4 §402. Now colour is essentially visible not in the first, but in this second manner, for visibility is a quality, as being snub is a quality of a nose. And this is why he says that colour is visible according to itself (*secundum se*) but not by definition (*non ratione*); that is to say, not because visibility is placed in its definition, but because it possesses of itself the reason why it should be visible, as a subject possesses in itself the reason for a quality proper to it.

n. 5 §403. Which he proves from this that all colour is able to move the diaphanous to act. For the diaphanous is the same as what is transparent—as air or water—and colour has this in its nature that it is able to move the diaphanous to act. And, on this, that it moves the diaphanous to act, the visible appears. Whence it follows that colour according to its nature is visible. And since the diaphanous is not brought to act save through light (*lumen*), it follows that colour is not visible without light. And, therefore, before it may be shown how colour may be seen, it is necessary to speak of light.

"The diaphanous is the same as what is transparent..." The diaphanous has in its nature something of the transparent and something of its contrary, the opaque. The transparent simpliciter (*aether*) is invisible and, likewise, the light it carries is invisible. Thus, the lights from the Sun and the stars do not manifest themselves in lighted pathways outside Earth's atmosphere. Their lights are only manifest on termination in the proper receptor, the eye (or its artificial equivalent, the photographic camera), or at the diaphanous (Earth's atmosphere), or at some *corporis terminatum* (bodily surface) whether outside Earth's atmosphere, such as a planet or satellite, or within it. In so far as the diaphanous is transparent it conveys light. It is in so far as it is a mix of the transparent and the opaque, it seems to me, that it makes colour manifest. It is the diaphanous, St Thomas says, that is receptive of colour. [Lectures XIV, n. 6; XV, n. 1]

n. 6 §404. Then, at 'There is, accordingly', he sets out those things without which colour cannot be seen, namely, the diaphanous and light (*lumen*); and this in three sections. First, he shows in what the diaphanous consists. Secondly, at 'But light is the act of this etc...' he sets out concerning light (*de lumine*) what is its act. Thirdly, at 'Now that only can receive colour', he shows how the diaphanous is receptive of colour.

To begin with, therefore, he says that since colour moves the diaphanous by its very nature, the diaphanous must clearly be something. Since the diaphanous does not have colour of its own, it enables things to be seen by receiving colour from outside, and in this peculiar fashion (*aliquo modo*) it is visible. Examples of the diaphanous are air and water and many solid bodies, certain jewels and glass. Now while other accidents pertain to the elements and the bodies of which they are constituted in accordance with the nature of those elements—such as heat and cold, weight and levity, and that sort of thing—the diaphanous does not befit the nature of air or water in this fashion (*tamen diaphanum non convenit praedictis*), but according to a common nature which is not confined to air and water—which are corruptible bodies—but to the heavenly body also which is perpetual and incorruptible. For at least some of the celestial bodies are manifestly diaphanous. We should not be able to see the fixed stars of the eighth sphere unless the lower spheres of the planets were transparent or diaphanous (*transparentes, vel diaphanae*). Hence it is evident that to be diaphanous (*diaphanitatis*) is not a property consequent on the nature of air or water, but of some more generic nature, in which the cause of diaphanousness is to be found, as we shall see later.

> "In order that it may enable vision, the diaphanous does not have colour of its own…" In fact the diaphanous (e.g., air, water, glass, diamond etc.) does manifest colour, albeit faintly, or very faintly. Both philosophers allow (cf., here, and in Lecture XV n. 2 below) that the diaphanous can be called visible in some respect.

> "[T]o be diaphanous is not a property consequent on the nature of air or water" but of some more generic nature. This is the issue. Aristotle uses one word to indicate the transparent, *diaphanes*. St Thomas uses two, *transparens* and *diaphanum*. What St Thomas is referring to here is not diaphanous-ness, but transparency, but he says *diaphanitatis* because he is unaware that the heavenly bodies— Sun, stars, planets, etc.—are not part of the heavenly matter. Transparency, I argue, can properly only be said of *aether*.

> Christopher Decaen offered the following advice on the view I have expressed here:

> "Note… that St Thomas also brings in the words *lucens* and *lucidus* and even *illuminans*, all referring to the light source, in chapters 14 and 15. It also occurs to me that St Thomas (esp. in *De Sensu*) sometimes uses *perspicuum* as a synonym for *diaphanum*. See, esp. ch. 5 of *De Sensu's Commentary*."[67]

n. 7 §405. Next, at 'Light (*lumen*) etc.', he shows what light (*lumen*) is, first stating the truth, then dismissing an error. To begin with he says that light (*lumen*) is the act of the diaphanous as such. For it is evident that neither air nor water nor anything of that sort is actually transparent (*transparens*) unless it is illuminated. Of itself the diaphanous is in potency to both light and darkness (the latter being a privation of light) as primary matter is in potency both to form and the privation of form. Now light (*lumen*) is to the diaphanous as colour is to a bodily surface (*ad corpus terminatum*): each is the act and form of that which receives it. And on this account he says that light (*lumen*) is the colour,

---

67   Personal communication to the author.

as it were, of the diaphanous, in virtue of which the diaphanous is made actually so by some light-giving body (*ab aliquo corpore lucente*) such as fire, or anything else of that kind, or by a celestial body. For to be full of light and to communicate it (*lucens actu et illuminativum*) is common to fire and to the celestial body, just as to be diaphanous (*esse diaphanum*) is common to air and water and to the celestial body.

> "[L]ight is to the diaphanous as colour is to a bodily surface... on [which] account [Aristotle] says that light is the colour, as it were, of the diaphanous..." Light is invisible in the transparent, as said above: it is visible only in the diaphanous.

> "[T]o be full of light and to communicate it is common to fire and to the celestial body, just as to be diaphanous (*esse diaphanum*) is common to air and water and to the celestial body." Here St Thomas elaborates his distinction of the diaphanous from the transparent. He speaks of the 'celestial body' in the two senses arising from the confusion with *aether*, the heavenly substance, of the celestial lights it seems to contain.[68] In the former, he applies to it the Latin word *lux*, indicating a light source; in the latter he is speaking of *aether* and its faculty of transparency by which *lux* is communicated.

n. 8 §406. Then, at 'We have then indicated' he rejects a false opinion on light (*de lumine*), and this in two stages. First, he shows that light (*lumen*) is not a body; then at 'Empedocles was wrong' he refutes an objection brought against the arguments which prove that light is not a body. As to the first point he does three things.

First, he states his own view saying that, once it is clear what the diaphanous is and what light (*lumen*) is, it is evident that light (*lumen*) is neither fire (as some have said positing three kinds of fire, the combustible, and flame, and light), nor a body at all, nor anything flowing from a body, as Democritus supposed, asserting that light (*lumen*) consisted of atomic particles emanating from a light giving body. If there were these emanations from bodies, they would themselves be bodies or something corporeal, and light (*lumen*) would thus be nothing other

---

68 'Confusion' is said here, not derogatively, but technically. Because of the limits of his experimental knowledge, St Thomas treats as one, elements which are physically distinct.

than fire, or something material of that sort, present in the diaphanous; which is the same as to say that light (*lumen*) is a body or an emanation from a body.

n. 9 §407. Next, at 'For it is impossible', he proves his own hypothesis thus. It is impossible for two bodies to be in one place at one time. If therefore light (*lumen*) were a body it could not co-exist with a diaphanous body; but this is false, therefore light (*lumen*) is not a body.

n. 10 §408. Thirdly, at 'But it seems' (i.e., 'Light seems') he shows that light (*lumen*) exists (*est*) together with the diaphanous. For contraries exist in one and the same subject. But light (*lumen*) and darkness are contraries in the manner in which privation and the possession (of a quality) is a species of contrariety as is stated in the *Metaphysics*, Book X [cf., chap. 4, 1055a30ff]. Obviously, darkness is a privation of this quality, i.e., of light (*lumen*) in the diaphanous, and therefore the diaphanous is the subject of darkness. Hence too, the presence of the quality mentioned, i.e., *lux*, is *lumen*: and therefore *lumen* exists (*est*) together with the diaphanous.

This passage elaborates St Thomas's understanding of light and the distinction he draws between *lux* and *lumen*, a distinction overlooked in standard translations which treat the two terms as synonymous. The subtlety of the Latin is not easy to render in English. I set it out here from 'But light and darkness...' to the end of the passage with the significant noun *habitus* (habit, power, quality or nature) highlighted. [L]umen autem et tenebra sunt contraria secundum modum quo privatio et **habitus** est quaedam contrarietas, ut dicitur in decimo metaphysicae. Manifestum est autem, quod tenebra est quaedam privatio huius **habitus**, scilicet luminis in diaphano; et sic subiectum tenebrae est diaphanum; ergo et praesentia dicti **habitus**, scilicet lucis, est lumen: ergo lumen est simul cum diaphano.

Where St Thomas first uses *habitus* he is referring to its usage in his Commentary on Chapter 4, Book X of Aristotle's *Metaphysics* [Book X, Lesson 6]. There it means 'possession' (i.e., 'something had', its nominal meaning). 'Possession' is there contrasted with 'privation' (*habitus* and *privatio*). The second time he uses it he is referring to the quality in the diaphanous whose privation is darkness, namely, *lumen*. But the third, and most significant usage, of *habitus* refers to the quality of which *lumen* is the representative in the diaphanous, *lux*. He will say, at n. 23 below, "the participation or effect of *lux* in the diaphanous is called *lumen*." He says there also that *lux* has no contrary, a consequence of its proper substance, *aether* ('first altering body') having no contrary. Here he says that *lumen* does have a contrary but only in respect of privation which, as is remarked in the passage in the Commentary on the *Metaphysics*, is a sort of contradiction (non-being) rather than contrariety *stricto sensu*. Light (*lumen*) is visible in the diaphanous, but invisible (as *lux*) in *aether*, as said above.

n. 11 §409. Then at 'And not rightly...' [i.e., 'Empedocles... was wrong'], he refutes an answer to one argument which might be urged against those who hold that light (*lumen*) is a body. For it is possible to argue thus against them: if light (*lumen*) were a body, illumination ought to be a local motion of light passing through the transparent; but no local movement of any body can be sudden or instantaneous; therefore, illumination would be not instantaneous but successive according to this view.

n. 12 §410. Of which the contrary is a fact of experience; for in the very instant in which a luminous body becomes present, the transparent is illuminated all at once, not part after part. So Empedocles, and all others of the same opinion, erred in saying that light was borne along by local motion, as a body is; and that it spread out successively through space, which is the medium between the Earth and its envelope, i.e. the sky; and that this successive motion escapes our observation, so that the whole of space seems to us to be illuminated simultaneously.

n. 13 §411. For this assertion is against the truth which reason can easily perceive. For the illumination of the diaphanous requires nothing other than the opposition to the body to be illuminated of the one illuminating with no obstacle intervening.

n. 14 §412. Again, it contradicts appearances. One might indeed allow that successive local motion over a small space could escape our notice; but that a successive movement of light from the eastern to the western horizon should escape our notice is so great an improbability as to appear quite impossible.

> It is worth repeating that modern science may criticise Aristotle and St Thomas on this point of instantaneous illumination, and say that they erred in rejecting Empedocles' view of light's successive motion. For light does not illumine instantaneously but successively and at a speed science can demonstrate, 299,792.458 km/s 'in vacuo'. But they were right and Empedocles, and modern science, wrong. For light is not a body, not corpuscular, not comprised of atomic particles, but a quality of a particular substance, *aether*. Crucial to understanding this is that light is not something that exists *in itself* (a substance) but only *in something else*. What follows? Light does not have a speed: rather, the speed of its propagation, *c*, is a property of its proper substance, *aether*. If the speed at which *aether* permits its propagation is not infinite, this is because *aether* is material and suffers from the limitations of all things material, a certain inertia.

n. 15 §413. But as the subject matter under discussion is threefold, i.e., the nature of light, and of the diaphanous, and of the necessity of light (*luminis*) for seeing, we must take these three questions one by one.

On the nature of light (*de natura luminis*) various opinions have been held. Some, as we have seen, held that light (*lumen*) was a body; being led to this by certain expressions used in speaking. For instance, we are accustomed to say that a ray 'passes through' the air, that it is 'thrown back', that rays 'intersect', and so forth; which all seem to imply something corporeal.

n. 16 §414. But this theory is groundless, as the arguments of Aristotle here adduced show, to which others might easily be added. Thus it is hard to see how a body could be suddenly multiplied over the whole hemisphere, or come into existence or vanish, as light does; nor how the mere intervention of an opaque body should extinguish light in any part of a transparent body if light itself were a body. To speak of the motion or rebounding of light is to use metaphors, as when we speak of heat 'proceeding into' things that are being heated or being 'thrown back' when it meets an obstacle.

n. 17 §415. Then there are those who maintain, on the contrary, that light (*lumen*) is spiritual [i.e., immaterial] in nature. Otherwise, they say, why should we use the term 'light' in speaking of intellectual things? For we say that intellectual things possess a certain intelligible 'light'. But this also is inadmissible.

n. 18 §416. For it is impossible that any spiritual or intelligible nature should fall within the apprehension of the senses; whose power, being essentially embodied, cannot acquire knowledge of any but bodily things. But if anyone should say that there is a spiritual 'light' other than the light that is sense-perceived, we need not quarrel with him; so long as he admits that the light which is perceived is not spiritual in nature. For there is no reason why quite different things should not have the same name.

n. 19 §417. The reason, in fact, why we employ 'light' and other words referring to vision in matters concerning the intellect is that the sense of sight has a special dignity; it is more spiritual and more subtle than any other sense. This is evident in two ways. First, from the object of sight. For objects fall under sight in virtue of properties which earthly bodies have in common with the heavenly bodies. On the other hand, touch is receptive of properties which are proper to the elements (such as heat and cold and the like); and taste and smell perceive properties that pertain to compound bodies, according as these are variously compounded of heat and cold, moisture and dryness. Sound, again, is due to local movement which, indeed, is also common to earthly and heavenly bodies, but which, in the case of the cause of sound is a different kind of movement from that of the heavenly bodies, according to the opinion of Aristotle. Hence, from the very nature of the object it would appear that sight is the highest of the senses; with hearing nearest to it, and the others more remote from its dignity.

n. 20 §418. Next, one can see how the sense of sight is more immaterial (*spiritualior*) from its mode of affectation. For in every other sense what is immaterial in its operation is accompanied by some natural change. I mean by 'natural change' what happens when a quality is received by a subject according to the material mode of the subject's own existence, as e.g., when anything is cooled, or heated, or moved about in space. But immaterial change (*immutatio spiritualis*) refers to the manner of reception of the likeness of an object in the sense-organ, or in the medium between object and organ, as a form, causing knowledge, and not merely as a form in matter. For there is a difference between the mode of being which a sensible form has in the senses and that which it has in the thing sensed.

Now in the case of touching and tasting (which is a kind of touching) it is clear that material change occurs: the organ itself grows hot or cold by contact with a hot or cold object—there is not merely an immaterial change (*non fit immutatio spiritualis tantum*). So too the exercise of smell involves a sort of vaporous exhalation; and that of sound involves movement in space. But seeing involves only an immaterial change (*immutatio spiritualis*), and hence among all the senses sight is the more immaterial (*spiritualior*); with hearing as the next in order. These two senses are therefore the most immaterial (*maxime spirituales*), and are the only ones under our control. Hence the use we make of what refers to them—and especially of what refers to sight, in speaking of intellectual objects and operations.[69]

N. 21 §419. Then again some have simply identified light (*lumen*) with the manifestation of colour. But this is patently untrue in the case of things that shine by night, their colour, nevertheless, remaining obscure.

N. 22 §420. Others, on the other hand, have said that light (*lumen*) was the substantial form of the Sun, and that the brightness proceeding therefrom (in the form of colours in the air) had the sort of being that belongs to objects causing knowledge as such. But both these propositions are false. The former, because no substantial form is in, and of itself, an object of sense perception; it can only be intellectually apprehended. And if it is said that what the sense sees in the Sun is not light but its splendour (*non est lux, sed splendor*), we need not dispute about names, provided only it be granted that what sight apprehends is not a substantial form. And the latter proposition too is false; because whatever simply has the being of a thing causing knowledge does not,

---

69   I have substituted 'immaterial' for St Thomas's 'spiritual' because in the 21st century we limit the use of the term 'spiritual' to matters which concern belief, or to the religious. In any event, 'immaterial' is just as effective in conveying his meaning.

as such, cause material change; but the rays from the heavenly bodies do in fact materially affect all things on Earth. Hence our own conclusion is that, just as the corporeal elements have certain active qualities through which they affect things materially, so light is the active quality of the heavenly body through which it acts; and is in the third species of quality, like heat.

n. 23 §421. But it differs from heat in this, that light (*lux*) is a quality of first altering body which has no contrary, whence it follows that light (*lux*) has no contrary: heat, on the other hand, has a contrary. And because light has no contrary there is no place for a contrary disposition in its recipient (*in suo susceptibili*). And, because of this, its matter (*suum passivum*), i.e., the diaphanous, is always as such immediately disposed to its form. That is why illumination occurs instantaneously, whereas what can become hot only becomes so by degrees. Now the participation, or effect, of light (*lux*) in the diaphanous is called *lumen*. If it appears in a direct line to the enlightened body it is called 'a ray'. But if it is caused by a reflection of a ray upon a light receiving body, it is called 'splendour'. But *lumen* is the universal [name] for every effect of light (*lux*) in the diaphanum.

> "If it appears in a direct line to the enlightened body it is called 'a ray'. But if it is caused by a reflection of a ray upon a light receiving body, it is called 'splendour'." Neither a ray of light, nor the splendour of light (as St Thomas defines it here) can occur in *aether*—which is not to say that the lights of Sun and stars seen from beyond Earth's atmosphere are not 'splendid'. But St Thomas is referring to that particular quality of light which accompanies its dispersal in the diaphanous.

n. 24 §422. So much being admitted as to the nature of light (*luminis*), we can easily understand why certain bodies are always actually lucent, whilst others are diaphanous, and others opaque. Because light (*lux*) is a quality of the first altering body, the most perfect and least material of bodies, those among other bodies which are the least material and most mobile are always actually lucent. The next in this order are the diaphanous; whilst those that are most material, being neither luminous of themselves nor receptive of light (*luminis receptiva*), are the opaque. One may see this in the elements. For fire has light (*lucem*) in its nature, though that light does not appear to us except in other natures on account of density. Air and water, being more material (*minus formalia*), are diaphanous; whilst Earth, the most material of all, is opaque.

> Here St Thomas expressly distinguishes *aether* ('first altering body') from the celestial bodies it appears to contain, and from the diaphanous, and ascribes *lux* to *aether* as its proper quality. He also ascribes *lux* to earthly fire.

n. 25 §423. As to the third point [the necessity of light for seeing], it should be noted that some have said that light is necessary for seeing on account of the colour in the things seen. For they say that colour has not the power to move the diaphanous, except through light (*nisi per lumen*). And they say that the indicator of this is that when one is standing in shadow he can see what is in the light (*in lumine*), but not conversely [i.e., if he stands in the light he cannot see what is in shadow]. The cause of this fact, they said, lay in a correspondence between sight and its object: as seeing is a single act, so it must bear on an object formally single; which would not be the case if colour were visible of itself—not in virtue of light—and light also were visible of itself.

n. 26 §424. Now this view is clearly contrary to what Aristotle says here, 'and which has in itself the cause of being visible'. Hence, according to Aristotle's opinion, it must be said that light is necessary for seeing, not because of colour, (as, they say, making colours actual which are only in potency while in darkness), but on account of the diaphanous which light renders actual, as the text states.

n. 27 §425. And as evidence of this, note that every form is, as such, a principle of effects resembling itself. Colour, being a form, has therefore of itself the power to impress its likeness on the medium. But note also that there is this difference between a form with a complete power to act and one with an incomplete power to act that the former is able not merely to impress its likeness on matter, but even to dispose matter to fit it for this likeness—which is beyond the power of the latter. Now the active power of colour is of the latter sort; for it is, in fact, only a kind of light somehow dimmed by admixture of opaque matter. Hence it lacks the power to render the medium fully disposed to receive colour. But this pure light (*lux pura*) can do.

n. 28 §426. Whence it is also clear that, as light (*lux*) is, in a certain way, the very substance of colour, all visible objects as such share in the same nature; nor does colour require to be made visible by extrinsic light (*per lumen extrinsecum*). That colours in light are visible to one standing in the shade is due to the medium's having been sufficiently illumined.

LECTURE XV

n. 1 §427. After the Philosopher has shown (above) what is colour, what is the diaphanous and what is *lumen*, he now proceeds to explain how the diaphanous is related to colour. It is clear, from the foregoing, that the diaphanous is receptive of colour; for colour acts upon it, as we have seen. Now what is receptive of colour must itself be colourless, as what receives sound must be soundless; for nothing receives what it already has. The diaphanous is therefore colourless.

n. 2 §428. But, as bodies are visible by their colours, the diaphanous must itself be invisible. Yet since one and the same power apprehends contrary qualities, it follows that sight, which apprehends light, also apprehends darkness. Hence, although the diaphanous of itself possesses neither light nor colour, being receptive of both, and is thus not of itself visible in the way that things bright or coloured are visible, it can, all the same, be called visible after the fashion of darkness which is hardly visible. The diaphanous, then, is of this sort, that is, darkness when it is not actually diaphanous, but only so in potency. For it is the same nature which is the subject at different times of darkness and of light (*lumen*). Thus it belongs to the diaphanous while ever it lacks luminosity and is only potentially transparent to be in state of darkness.

> "[A]s bodies are visible by their colours, the diaphanous must itself be invisible…" But relatively, not absolutely, so because the diaphanous is receptive of both light and colour. Both philosophers agree that the diaphanous can be called visible after the fashion of darkness and the scarcely visible. And this is borne out by experience, for each instance of the diaphanous, e.g., air, water, glass and (clear) precious stones, is coloured, albeit faintly. Similarly, considered as the media of sound, neither air nor the sea is utterly soundless as, e.g., when either is agitated.

n. 3 §429. Then at 'Not all', having decided about colour, which is made visible by light, he reaches a conclusion about that other visible object of which he said above that it had no proper name. He observes that not all things depend on light for being seen, but only the colour that is proper to each particular thing. Some things, e.g., certain animals that appear fiery and lucent in the dark, are not visible in the light, but only in darkness. There are many such things, including the fungi of oaks, the horn of certain beasts and heads of certain fish, and some animals' scales and eyes. But while all these things are visible in the dark, the colour proper to each is not seen in the dark. The things are seen both in light and in darkness; in light as coloured objects, but in darkness only as bright objects.

n. 4 §430. The reason why they are seen shining in the darkness is another matter. Aristotle only mentions the fact incidentally, in order to show the relation of the visible to luminosity. This, however, seems to be the reason for their being visible in the dark, that such things have in their constitution something of light (*aliquid lucis*), inasmuch as the brightness of fire and the transparency of air and water is not entirely smothered in them by the opacity of Earth. But having only a small amount of light (*modicum... de luce*), their brightness (*lux*) is obscured in the presence of a greater light (*maioris luminis*). Hence in the light they appear not as bright, but only as coloured. But their light is so weak that it is unable perfectly to actualise the diaphanous so as to reduce it perfectly to act so that it can bring forth colour. Hence, by their light (*sub eorum luce*) neither their colour, nor that of other things, is able to be seen: only their brightness (*lux*). For light (*lux*), being a more effective agent upon the diaphanous than colour, and more visible, can be seen with less alteration of the diaphanous.

n. 5 §431. Next, at 'But now' [i.e., at 'At present what is clear...'], he explains how colour actually affects sight, first pointing out what this necessarily presupposes; and then, at 'The same holds', he shows that something similar necessarily applies in respect of the other senses. Concerning the first he makes two points. First he establishes the truth. Then at 'This is, however, impossible' he excludes an error. First, then, he says, what is clear as mentioned above, that what is seen in light (*in lumine*) and cannot be seen without it, is colour, for as said above, it is of the nature of colour (*de ratione coloris*) to move the diaphanous; and it does this through light (*lumen*) which is the act of the diaphanous. Therefore without light (*lumen*) colour cannot be seen.

> "[W]ithout *lumen* colour cannot be seen." St Thomas's careful distinction of *lumen* from *lux* here is crucial to the understanding of how the diaphanous differs from *aether*. Colour cannot appear unless there is *lumen*—not *lux*, be it noted, but *lumen*—which can only appear in the diaphanous. The applications of modern science bear this out: photographs of objects which have no surrounding atmosphere show them to be practically devoid of colour. Photographs of objects in the presence of an atmosphere, as e.g., the surface of Mars, do manifest colour.

n. 6 §432. The sign of which is this: if a coloured body is placed upon the organ of sight it cannot be seen, for there is no diaphanous medium to be affected by the colour. For though the pupil is [itself] a sort of diaphanum, yet it is not diaphanous in act if the coloured body is placed upon it. For there has to be air or something of that sort for colour to move the diaphanous to act, by which the [operative] sense, that is, the organ of sight, is moved as by a body continuous with itself. For bodies only affect one another through contact.

n. 7 §433. Then, when he says 'For this (is impossible)' he sets aside an error, saying Democritus did not speak well in opining that if the medium between the eye and the thing seen were a vacuum, any object, however small, would be visible at any distance, e.g., an ant on the vault of heaven. This is impossible. For if anything is to be seen it must actually affect the organ of sight. Now it has been shown that this organ as such is not affected by an immediate object—such as an object placed upon the eye. So there must be a medium between organ and object. But a vacuum is not a medium; it cannot receive or transmit effects from the object. Hence through a vacuum nothing would be seen at all.

n. 8 §434. Democritus erred because he thought that the reason why distance diminishes visibility was that the medium is an impediment to the action of the visible object upon sight. But this is false. The diaphanous is not in the least incompatible with luminosity or colour; on the contrary, it is precisely proportioned to their reception; a sign of which is that it is illuminated or coloured instantaneously. The reason why distance diminishes visibility, is that everything seen is seen within the angle of a triangle, or rather pyramid, whose base is the object seen and apex in the eye that sees.

n. 9 §435. It makes no difference whether seeing takes place by a movement from the eye outwards, so that the lines enclosing the triangle or pyramid run from the eye to the object, or the opposite, so long as seeing does involve this triangular or pyramidal figure; which is necessary because, since the object is larger than the pupil of the eye, its effect upon the medium has to be scaled down gradually until it reaches the eye. And, obviously, the longer are the sides of a triangle or pyramid the smaller is the angle at the apex, provided that the base remains the same. The further away, then, is the object, the less does it appear—until at a certain distance it cannot be seen at all.

n. 10 §436. Next, at 'But fire (is seen)', he explains how fire and bright bodies are seen—which are visible not only, like coloured objects, in the light, but even in the dark. There is a necessary reason for this, namely that fire contains enough light to actualise perfectly the diaphanous, so that both itself and other things become visible. Nor does its light fade in the presence of a greater light, as does that of the objects mentioned above.

4. GRAVITY

In principio creavit Deus caelum et terram...

Genesis I: 1

*"[T]he entire universe is to be considered prior to its
parts, simple bodies before the compound, [and] among
simple bodies the first, the heavenly body through which
all others are sustained, is first to be considered..."*

St Thomas Aquinas[70]

When Sir Isaac Newton propounded his formula for the universal law
of gravitation, he did not make the mistake of confusing its calculation with its causation.[71] Criticised for arguing to the existence of some
external and undetectable cause, he responded that it was enough that
phenomena implied attraction but he had never "[sought to assign] a
cause to this power".[72] In his letters to Richard Bentley he added this
gloss:

> "Gravity must be caused by an agent acting constantly according
> to certain laws. But whether the agent be material or immaterial, I
> have left to the consideration of my readers."[73]

He knew, despite his embrace of a qualified materialism, that space could
not be a void; but the same materialism misled him as to the nature of
the body that filled the universe. Newton's successors disregarded his

---

70   *In I De Caelo*, Prologue.

71   Cf. http://en.wikipedia.org/wiki/
Newton%27s_theory_of_gravitation#Newton.27s_reservations

72   *Principia Mathematica* Bk. III, General Scholium. "I have not been able to discover
the cause of [the] properties of gravity from phenomena, and I frame no hypotheses... [I]t is
enough that gravity... acts according to the laws which we have explained..."

73   In this sentence he shows he is open to the suggestion that something might be real
yet not comprised of matter. His successors mocked him for his limited adherence to
metaphysical principle.

caveat and treated gravity's causation as identical with its calculation. They did more: they rejected his assertion that gravity had an instrumental efficient cause and, by implication, the need for a principal efficient cause. The protocols of materialism led them to embrace a practical atheism.

Einstein embraced the materialistic paradigm thoroughly as he immersed himself in the thought of Hume and Mach. He accepted the materialist conclusion flowing from the Michelson-Morley experiment that no ether existed. Space seemed, from observations, to exercise a certain causative faculty. Uninhibited as Newton had been by a residual metaphysics, he saw no difficulty in ascribing such causality to something which, on any assessment, was bereft of any objective reality at all! Gravity was a natural outcome, he said, of the presence of the mass of a body in space. It 'warped' the space around it, impelling other bodies, should they approach too close, to depart from their rectilinear paths. The greater a body's mass, the more it 'warped' the space around it. Gravity was not a force propagated between bodies but the inevitable effect of the interplay of their mass and the surrounding space.

It is appropriate to repeat the problems with these materialistic views of gravity's causation.

1. There is nothing in a body, *qua* body, which requires that it should attract another.

2. Any assertion of causes which assumes that space is a void, is grounded in an impossibility.

3. If space is a void, this "non-being somehow existing" would present an absolute barrier to transmission of gravitational force, as it would to the transmission of light.

4. If space is a void, a logical dilemma follows. Einstein's theories hold that the speed of gravity's propagation is determined at c, "the speed of light", 299,792,458 metres per second. But if space is "non-being-somehow-existing" no reason can be advanced why the speed of gravity's propagation, or that of light, is not infinite.

5. If space is a void, there is no medium whereby the immense forces of attraction demanded by treating its causation as identical with Newton's calculations, or the 'natural' inclination

of space under the influence of mass posited by Einstein, can be conveyed—the dilemma of action at a distance.

6. Neither the explanation of Newton's successors nor that of Einstein provides an adequate account for the effect, as universal as is gravity, of the sphericity of form of celestial bodies.

7. Neither explanation provides any account at all for the effect, equally universal among the heavenly bodies, of circular motion.

8. Each explanation supposes a metaphysical impossibility, the absence of an efficient (*extrinsic*) cause. Newton was prepared to allow one. The need for one never entered Einstein's head.

For all the sophistication of its knowledge, modern science's explanations as to how gravity operates are irrational. It has yet to discover gravity's cause.

The many attempts to explain gravity as a species of extrinsic force have foundered over difficulties about the nature of the force and the mode of its operation. In the late 1740s Georges Louis Le Sage, for example, proposed a mechanical explanation arguing that gravity's force was constituted by particles of great rarefaction.[74] But science revealed that material bodies are largely porous, 'mostly empty space'. His hypothetical particles would be expected, then, to penetrate, rather than to bear upon, the surface of celestial bodies.

The philosopher Immanuel Kant raised a more fundamental objection. Le Sage's particles must, he argued, have a 'non-zero' radius. This implied the existence of some sort of binding force to hold these particles together. Now, *that* binding force could not be explained by the gravitational particles themselves. Hence there had to be some additional force binding these, and so on, *ad infinitum*. This objection, addressing as it did the influence which provides extension and parts to a material substance, the metaphysical category *quantity*, demonstrated that one could not hope to discover an extrinsic cause of gravity and ignore the force that binds atomic and molecular structure. It also showed that a substance that could produce such a force must be superior to any ordinary material substance.

---

74   Cf. http:en.wikipedia.org/wiki/Le_Sage%27s_theory_of_gravitation

* * *

Modern science is concerned with things observable, with phenomena. Its theories, its determinations, its prognostications, are all grounded in mensurable data. It notes effects, it looks for causes to explain them; and because its *modus* is almost exclusively inductive the causes at which it arrives are not necessarily the true causes or, if true, are not necessarily the ultimate causes. For certitude in induction depends upon the discovery of a sufficiency of effects to exclude error about the cause. An explanation may 'save the appearances', as St Thomas remarked about Ptolemaic astronomy, and not exclude the possibility of another theory providing a better.[75]

Modern science has another limitation. It is informed—and has been for some 350 years—by a defective philosophy manifest in two poles of thought, materialism and subjectivism. The one contends that if something cannot be detected experimentally, it does not exist; the other that only that is true which the individual, or the majority, asserts to be true. These defects reflect the mentality of worldly thinkers who have long since turned their backs on any philosophy which addresses the part of reality which is not material. The result is that experimental science frequently fails to reach sound conclusions. Has there ever in the history of mankind been an age to compare with the present in breadth of knowledge, and lack of wisdom? T. S. Eliot put it succinctly:

> Where is the wisdom we have lost in knowledge?
> Where is the knowledge we have lost in information?
> Endless invention, endless experiment
> Brings knowledge of motion, not of stillness...
> The cycles of heaven in twenty centuries
> Bring us farther from God and nearer to the Dust..."[76]

* * *

---

75   *Summa Theologiae*, I, q. 32, art. 1, ad 2.
76   *The Rock*

The modern world is apt to dismiss the thinking of Aristotle and St Thomas Aquinas in the realm of nature because it can show their cosmology to be defective. But their cosmology was not so much defective as limited. Their analyses, grounded in being rather than its accidents and the limited vision of modern philosophy, more than compensate for their shortcomings in knowledge. I propose to set out the principles they expounded concerning the behaviour of the heavens and gravity—though neither recognised this latter as the entity whose laws were codified by Newton—and to revisit their insistence on the existence in the natural world of an element which modern science refuses to acknowledge.

Let us begin with a self-evident principle: *nothing, i.e., non-being, does not exist*. Its corollary is this: every material thing is surrounded by other material being however intangible. We may accept this readily enough in respect of the bodies we encounter in daily experience. Even if we cannot discern, we can imagine the proximity (taken literally here as 'the next-ness') of other material being to the very least of bodies: but what of the celestial bodies in 'outer space'? What of the atoms and molecules of which material bodies are constituted but surrounded, so science tells us, by 'empty space'? Neither of these 'spaces' can be empty: principle prevents it.

Modern science maintains that notwithstanding that light is a material reality it does not need a material medium in which to travel. How can this be? If light encountered a somehow existing 'nothing', this 'nothing' would be an impenetrable barrier to its passage.[77] Einstein tells us that the relationship between energy and the mass of a body is a function of the speed of light. But what has that ethereal, if powerful, reality, light, to do with the relationship between those two? The answers to these questions lie in the acknowledgement of the existence—and the remarkable characteristics—of an element of the material universe whose reality was exposed by Aristotle. He called it 'the heavenly body', or 'first body', or *aether*; later thinkers have referred to it, perhaps dismissively, as 'the quintessence'.[78]

---

77   On the absolute impediment of void, if it did exist, to the passage of any material thing see Aristotle, *Physics*, Bk. IV, vi (213b 30 et seq.) and St Thomas's Commentary, *In IV Physics* L. 10.

78   Literally, the fifth essence. The four material essences of philosophical antiquity were *earth, air, fire* and *water*. *Aether* was the fifth.

What follows are the relevant teachings of Aristotle in natural philosophy and cosmology with comments by St Thomas Aquinas. I will refer to the two hereafter collectively as 'the philosophers'. For the purposes of this exercise the reader is asked to accept, for the moment, the limitations in their knowledge. The references for the most part are from St Thomas's commentaries.

The Principles enunciated by Aristotle and St Thomas

i. Anything that moves is moved by another.[79]

ii. Nature is the principle of motion in all moveable things, in two ways:

> "[O]ne is active, i.e., the mover, as the soul is the active principle of the movement of animals; the other [principle] is passive, according to which a body is apt to be moved. Such... [is present] in the heavy and the light, for these are not composed of a mover and a moved, for... it is plain that none of these moves itself; each has with respect to its motion, a principle not of acting but of being acted upon." [*In I De Caelo*, L iii, 22]

iii. The *passive* principle of the motion of the heavens is that body's nature according to which it is apt to be moved with such a motion, but:

> "the active principle of the motion of the heavens is an intellectual substance..." [*In I De Caelo*, L iii, 22]

iv. There are four species of motion—generation, increase, alteration and local motion [*In IV Physics* L23, 631]—but the first, the more simple, and regular of motions is this last, local motion. [*In VIII Physics* L 14, 1094-5] All local motion is either straight (rectilinear) or circular,

---

79 *Physics* Bk. VIII. This is clear for things inanimate; not so clear for the animate differentiated precisely in the fact that they do move themselves. These have an interior principle, a soul, which causes their movement (whether as to *execution* only (plants), as to *execution* and *form* (brute animals), or as to *execution*, *form* and *end* (rational animals). But even the animate, Aristotle shows, is ultimately moved by another. This process of being moved cannot proceed to infinity: there must, therefore, be a first unmoved mover of all other beings.

or some combination of these two. [*In VIII Physics* L 16, 1105] Straight motion is imperfect because it involves contraries; for it must cease when it reaches its term, or return by reflex motion to its beginning. [*In VIII Physics* L 16, 1106] In contrast, circular motion:

> "is more simple and perfect... [It] is not corrupted when it reaches the terminus (since its beginning and end are the same)... The perfect, moreover, is prior to the imperfect... in nature, in *ratio* and in time... Circular motion, therefore, must be prior to straight motion." [Cf. *In VIII Physics*, Ll. 14-19; this from l. 19 towards the end.]

v. The universe is spherical and, since all motion is founded upon something immobile [*In I De Caelo*, L iii, 36], its motion must be considered in relation to its immobile centre. Hence, reflecting their respective relations to the universe's centre, there can be only three simple natural motions—one *from* the centre, one *towards* the centre, and a third *around* the centre. [*In I De Caelo*, L iii, 36]

vi. It is impossible that the heavens be comprised of a void:

> "[F]or there is no such thing as a self-existing void." [*Physics* IV, 8 (216 a & b)]

vii.   A simple body is one that has a principle of natural motion. There are four simple elements: earth, air, fire and water. Fire and air have a principle of straight motion *away from* the centre (of the universe), as earth and water have a principle of motion *towards* its centre. Circular motion is perfect motion; it has no contrary. Such motion, simple and distinct from straight motion, must be proper to some natural simple body other than these four. [*In I De Caelo*, L iii, 36]

> "[F]or the contrary of one thing (under the same respect) is one [*Metaphysics* X] and the motion contrary to an upward motion is a downward one. Hence, circular motion cannot be its contrary... [*In I De Caelo*, L iii, 38]

"Prior motion naturally belongs to a prior body. Now straight motion naturally belongs to some one or other of the [four] simple bodies... And if it happens that straight motion is found in mixed bodies [bodies comprised of two or more of the four simple elements] that will be due to the nature of the simple body predominant in it. As a simple body is naturally prior to the mixed, so circular motion is proper, and natural, to some simple body which is prior to the elementary bodies that exist here among us." [*In I De Caelo*, L iv, 41]

viii. This fifth element [*quintessence*], as befits a substance with perfect movement, is perfect. It is higher and nobler than the four simple elements. It is incapable of being generated or corrupted; incapable of expulsion from its proper place by violence; it has no lightness or heaviness; it has no contrary; it is ontologically prior to, and contains, all other bodies.

"[S]ince motion is proportionate to the mobile as is act [to its potency], it is fitting that a body which is un-generable and incorruptible and incapable of expulsion from its proper place by violence should have circular motion... [*In I De Caelo*, L iv, 38]

"[I]n order for something to be partially perfect it must have the beginning, middle and end in itself; but to be completely perfect it is required that there be nothing outside it. And this mode of perfection belongs to the first and supreme body which contains all bodies..." [*In I De Caelo*, L iv, 42]

ix.   This element moves other bodies.

"[Aristotle's] fourth argument proceeds from two assumptions. The first is that every simple motion is either according to nature or outside nature. The second is that a motion which is outside nature for one body is according to nature for another... Now it is manifest that circular motion is present in some body which the senses observe to be moved circularly. And if such a motion is natural to it... there [must] be an additional body which is moved circularly. But if circular motion is outside the nature of the body so moved, it follows from the foregoing assumption that for some other body it is according to nature and this body will be of a different nature from the four elements." [*In I De Caelo*, L iv, 46]

"Whatever is present in lower bodies from the impression of a higher is not violent or against nature, for they are naturally apt to be moved by the higher body." [*In I De Caelo*, L iv, [39]]

Moreover, as it contains all other bodies, this element is to them as form to matter and as act to potency. [*In I De Caelo*, L iv, 50]

x. But this element cannot be moved by other bodies as Aristotle teaches:

"While usually the thing touching is touched by what it touches... still it also occurs... that only the mover may touch the moved, while the thing touched does not touch the one touching it... [*De Generatione et Corruptione*, Bk 1, Pt. 6][80]

And St Thomas, commenting on Aristotle's *Physics*, remarks:

"Bodies act upon each other by touching... But this should be understood [only] when there is mutual contact as happens in those things that share a common matter... The heavenly bodies, however, because they do not share a common matter with inferior bodies, act upon them in such wise that they are not acted upon by them; they touch and are not touched." [*In III Physics*, L. 4, n. 5][81]

xi. This, the heavenly body, which St Thomas refers to as 'first altering body', is universal in the heavens. Following Plato, Aristotle and St Thomas we will refer to it as *aether*.

---

80   Cf. Christopher A Decaen in *Aristotle's Aether and Contemporary Science*, op. cit., footnote 50.

81   Cf. Christopher A Decaen in *Aristotle's Aether and Contemporary Science*, op. cit., footnote 51. Apparently St Thomas did not comment on Bk. 1, Pt. 6 of Aristotle's *De Generatione et Corruptione*. Note that here St Thomas includes with the heavenly substance, *aether*, the celestial bodies, Sun, Moon, stars and planets which seemed to be part of it.

Here, in summary, are the characteristics of *aether* the philosophers expose:

a. It is moved by an intellectual substance;

b. Its proper motion is perfect, i.e., circular, motion;

c. It is a simple natural body distinct from the four simple natural bodies, earth, air, fire and water, and from any body comprised of two or more of these;

d. It is perfect, higher and nobler than other simple elements;

e. It is incapable of generation or corruption;

f. It is incapable of expulsion from its proper place by violence;

g. It has no lightness or heaviness;

h. It has no contrary;

i. It is ontologically prior to all other bodies;

j. It contains all other bodies;

k. It is to all other bodies as form is to matter and as act is to potency;

l. It moves other bodies;[82]

m. But cannot be moved by them.

To this list must be added the philosophers' insistence that all motion in the universe is founded on something immobile.

---

82   Which must be understood rightly. Ultimately, every material thing, even the automotive (the living), is moved by the First Mover, God. Things are moved immediately by instrumental causes and under different respects; moved in as many ways as there are species of motion (cf. II, iv above). *Aether* is such an instrumental cause, perhaps the most fundamental in the material order.

Applying these Principles to the Facts exposed by Modern Science

i. The Earth is a globe turning on its axis once each day. It circles the Sun each year, its rotational axis inclined some 23 degrees to the perpendicular of the plane of its orbital axis. The Moon is a satellite circling the Earth every 27.3 days, though taken with respect to the Earth's motion around the Sun it takes 29.5 days for it to return to the same phase. These heavenly bodies and all the stars and planets that people the sky are immersed in an apparently empty sea of space. Sun, Moon and stars do not rotate around the Earth each day. Their apparent daily circuit is a function of the planet's rotation. The elements of which material bodies are comprised are not four but (at last count) 118.[83]

ii. The very first words of Divine revelation are: *In the beginning God created the heaven and the earth...* Notice that the sacred author first says that God created the heaven. This accords with philosophical principle. No material thing can come into being, can exist, *save in a pre-existing material setting*. The Earth and its various component parts—the stars, the Sun, the celestial bodies, and the elements of which these are comprised—none could have been created *in vacuo*; for a self-existing void—non-being somehow existing—is impossible. What, then, are we to understand by the expression 'the heaven' in the text? It is not unreasonable to conclude that it signifies not the Sun, Moon, planets, stars and other celestial bodies, but *aether* 'the heavenly body'. What are we to understand by the expression 'the earth'? It is not unreasonable to hold that it signifies not just our own planet but all celestial bodies, indeed, all instances of ordinary, or common, material being in the universe.

When the philosophers say that *aether* is 'first body' they mean exactly that; *first* in the order of reality, *first* in the order of time. The apparently empty void of space is replete with the first body. But let it be understood that 'replete with' is not convertible with 'filled' as if *aether* was a fluid poured into an empty vessel. *Aether*, not void, is first

---

83 One critic has suggested that the assertion of the fourfold constitution of the elements of the material world accepted by the ancient Greek philosophers is reflected in the modern division of material structures into solids, liquids, the gaseous and plasma.

in the order of reality: where there is no other material being, there is *aether*. While imagination inclines us to view a void as reality's 'default setting' (to adopt modern computer jargon), intellect insists it is the first body, *aether*. *Aether is* the universe, the matrix in which every celestial body exists; the universe is *aether*. St Thomas appears to acknowledge this in commentary (second lecture) on the first book of the *De Caelo*.[84]

iii. However, when it is said that *aether* is an element of the natural world this is not to be understood in the sense of 'a component'. St Thomas writes:

> "It is to be noted that Aristotle here reckons the heaven [*aether*] among the elements, although an element is something out of which things are composed, as said in *Metaphysics* V. However, while [*aether*] does not enter into the composition of mixed bodies it is involved in the composition of the whole universe as being part of it. Either that, or he is using the word 'element' in a wide sense to designate any of the simple bodies... to distinguish them from prime matter..."[85]

Now what is said about *aether's* involvement in the operations of the celestial bodies is, in my view, no less true of the substances of which they are comprised. These do not subsist, do not come into existence, save in *aether* as their proper matrix. Much as the sea is the medium and essential condition in which fish and other creatures exist, *aether* is the medium and essential condition of the existence and coming into existence (the '*be*' and '*become*') of all material things. Accordingly, *aether* cooperates with first metaphysical accident *quantity* in binding the atomic and molecular structure of each bodily substance.

iv. The philosophers conceived of the celestial bodies as embedded in concentric spheres with their motions determined by this heavenly body. They were aware of the reality of gravity as 'heaviness' and that this involved a force or tendency downwards. They understood, too, that the Earth was spherical, but held it to be fixed and its centre the

---

84 *In I De Caelo*, L 2, 17.
85 *In I De Caelo*, L. xviii, n. 7

centre of the universe. If certain bodies had a downward motion it was because that was part of their nature.

Modern science notes a number of effects universal among celestial bodies—spherical formation; circular movement; gravity. We know that the Moon rotates about the Earth; that the planet circles the Sun; that the Sun and stars circle in our own galaxy, the Milky Way, and that stars circle in other galaxies.[86] If science has confirmed anything it is that circular motion is as characteristic of the bodies that people the universe as it is uncharacteristic of the mundane. Now this universal effect, as with those of spherical form and gravity, must have a proportionate cause.

v. Many realities in nature are not scientifically detectable. Science cannot, for instance, detect experimentally the cause of life in a living being. This reality has no weight, no colour, no appearances, nothing which can be measured, because it is not a material reality but an immaterial one. One can only conclude to its existence from effects. Yet it is the essential element of the living thing.[87] Notwithstanding its besotted-ness with materialism science cannot, accordingly, object to some reality simply because it is unable to discern the presence of, or measure, its physical characteristics. Science's bemusement over entities it postulates, such as 'dark matter' and 'dark energy', exposes the defects of the philosophy to which it adheres.

Now cause and effect are always proportionate. The more particular an effect, the more particular is its cause; the more universal an effect the more universal its cause. If the thesis proposed here be accepted, it is clear that the only substance as universal as the effects of gravity, spherical form, and circular movement, is *aether*. As a working hypothesis, then, let us assume that *aether*, the sea in which according to this thesis all celestial bodies subsist, is the cause of these effects.

86   Notably M31, Andromeda Galaxy, and M33, Triangulum Galaxy. Those who contend that the motion of the Earth and of other planets and satellites is not circular but elliptical are splitting hairs. Whatever the effects of the modifying influences, the motion is primarily and *per se* circular.

87   For if it is lost the thing ceases to be. For living things, Aristotle teaches, to live is the same as to be.

vi. Let us recall what the philosophers have to say of philosophical principle and the attributes they ascribe to this element, and weigh these against the realities science cannot explain.

- *All motion is founded on something immobile.*

- *Nothing moves that is not moved by another.*

- *Nature is the principle of motion in all moveable things.*

- *Two principles underlie all motion, the one of acting, the other of being acted on.*

Their contention that the immobile thing on which the mobility of the universe was founded was the physical centre of what they perceived to be its sphere is now shown to be problematic.[88] In perceiving the heavens (i.e., *aether*) as in motion they were, of course, misled by the limitedness of their knowledge derived from the rotation of the planet on its axis; that is, they were misled by appearances.

In the Michelson-Morley experiment (1887), science demonstrated that *aether* is undetectable. While it is certain that *aether* moves *per accidens,* for it adapts to the movement of all other bodies, *per se* it would seem to be immobile. Indeed, careful reflection on the curious nature of this substance indicates that *aether* is immobile with respect to every element of common material being. Hence, even with respect to heavenly bodies moving at great speed in opposition to each other, the

---

88   Though for each celestial body there remains 'a still point of the turning world', to quote T S Eliot [*The Four Quartets*]. While no such body is absolutely immobile due to the influence, as science perceives it, of other celestial bodies, each such body is yet relatively so. St Thomas sheds light on the issue where he deals with an objection to Aristotle's view that circular motion is a simple motion. "[T]he parts of a sphere which is in circular motion are not in uniform motion but the parts near the poles or near the centre are moved more slowly because they traverse a smaller circle in a given time; consequently the motion of a sphere seems to be composed of fast and slow motions." His answer is instructive: "But it must be said that a continuum does not have parts in act, only in potency. Now what is not in act is not in actual motion. Hence the parts of a sphere, since they form a continuous body, are not actually being moved. Hence it does not follow that in spherical or circular motion there is diversity actually, only potentially." [*In I De Caelo*, L iii, 26]. Transported on the surface of the Earth at a rotational speed of 465 metres per second (at the equator), we are quite unconscious of the motion. One critic has advanced against this unconsciousness that the reason is that the Earth's motion is uniform, not bumpy. Yet 'bumpy' is said relatively to some standard; but there is no standard against which Earth's motion can be measured.

aether in which each subsists is immobile.[89] In summary I contend that *aether* is *the immobile 'something'* on which the motion of the heavenly bodies is founded and, to this extent, I would depart from the philosophers' teaching that circular motion is *aether's* proper motion and argue, instead, that circular motion is the proper effect *aether* induces in these bodies.

vii.   In line with this thesis *aether* is the principle of acting, the active principle, of heavenly motion but as instrument, not as principal. For the principal active principle is, as the philosophers teach, an intellectual substance. The passive principle of the motion of the celestial bodies is embodied in their nature as elements of common material being. Consistent with this is St Thomas's assessment that, since it contains all other bodies, *aether* the heavenly body, is to them as form is to matter, as act is to potency; i.e., it is their determinant. *In I De Caelo*, L iv, 50

Had the philosophers known of the discoveries of modern cosmology and those of science concerning the elements and their periodic table they would have had no difficulty adapting their teaching to encompass 118, rather than four, elements to maintain that the quintessential body to whose existence they had concluded was:

> "of a different nature from [those]... elements." *In I De Caelo*, L iv [46]

For the motion proper to each of the 118 elements, as to the almost infinite number of their compounds, is rectilinear motion, and:

> "if circular motion is outside the nature of a body that is moved circularly... for some other body it is according to nature..." *In I De Caelo*, L iv, 46

This last requirement is satisfied, in my view if, rather than being *aether's* proper motion, circular motion is *aether's* proper effect 'according to nature'.

---

89   This characteristic offers an explanation for the fixity of *c*, the speed of light. It reflects the fixity of its proper substance.

viii. Therefore the motion of each celestial body about its own axis, as of its movement about another body or bodies, is governed by a substance whose proper effect according to nature is the induction of circular motion in other bodies, a substance different from, and superior to, those other bodies; that is, Aristotle's 'heavenly body', *aether*.

How does Aether operate?

"What put you on to this...?"

"Aristotle chiefly... He says, you know, that one should always prefer the probable impossible to the improbable possible."
Lord Peter Wimsey[90]

i. Gravity is an accident. It exists in substances such as the heavenly bodies and their component parts. As with every other created reality gravity has four causes; two are intrinsic to the effect (the *material* and the *formal* causes) and two extrinsic (the *efficient* and *final* causes). Its *formal* cause is the inclination of the parts of the globe whether actual or potential (i.e., above, on or within it) of which a celestial body is comprised towards its centre: the *material* cause is the globe and its parts. The *final* cause, taken as the *ultimate* end of the operation, is the ordering of the globe and its parts to the good of the whole. Taken as the *immediate* end of the operation it is its focus or centre, the centre of gravity of the body and, with bodies in proximity, their common centre of gravity or barycentre. The *efficient* cause is the agent that produces the gravitational effect. It is twofold, *principal* and *instrumental*. The *principal* efficient cause is an intellectual substance, the *instrumental* cause is the means the principal uses.

Sir Isaac Newton may have forsaken the teaching of Aristotle for that of Descartes but he had not forsaken metaphysical principle so far as deny the rational claim for the need of an (*extrinsic*) efficient cause of the phenomenon he was studying. Here is the challenge he left for those who would follow him:

90   Dorothy L. Sayers' fictional detective in *The Unpleasantness at the Bellona Club*, London (Victor Gollancz Ltd.), 1921, Ch. XV. This (admittedly flippant) citation is from Aristotle's *Poetics* (cf. Bekker 1460a) where the context is human making (the artificial) rather than that of the Divine Author (the natural).

"It is inconceivable that inanimate Matter should, without the Mediation of something else, which is not material, operate upon, and affect other matter without mutual Contact... That Gravity should be innate, inherent and essential to Matter, so that one body may act upon another at a distance thro' a Vacuum, without the Mediation of anything else, by and through which their Action and Force may be conveyed from one to another, is to me so great an Absurdity that I believe no Man who has in philosophical Matters a competent Faculty of thinking can ever fall into it. Gravity must be caused by an Agent acting constantly according to certain laws; but whether this Agent be material or immaterial, I have left to the Consideration of my readers."[91]

Clearly the *instrumental* efficient cause of gravity cannot be the globe itself for the efficient cause is always *extrinsic* to the effect it produces and the globe and its parts, because they are the matter of the effect, are part of its *intrinsic* causes. It is impossible, moreover, that the same can, under the same respect, be both intrinsic and extrinsic causes of the same effect.[92] Just as clearly, the cause must be an agent which is as universal as is the effect.

ii.   Aristotle remarks on the contrariety to be found in nature.[93] There should be no surprise, then, that the superiority of *aether* to bodies of ordinary matter involves contrariety in its action. A body of ordinary matter acts *from within* for nature is an intrinsic principle providing determinate powers, acts and ends to its subjects. But *aether*, as befits the container of all other bodies, acts *from without*. It is extrinsic to

---

91   Letters to Dr Richard Bentley, 1692-3.

92   It may be objected that magnetic attraction between two bodies of iron involves an *intrinsic* efficient cause but it is not so. The efficient cause of magnetic attraction is the Author who invested it with that inherent property. An analogy may be drawn from the case of a rifle whose inventor (extrinsic to it) is the efficient cause of the inherent property that enables it, when operated, to emit a projectile at great speed.

93   The constitution of a body involves two contrary principles, one of being determined (prime matter), the other of determining (substantial form). St Thomas addresses the issue directly in the citation in II, ix above when he says this element is to all other bodies as form is to matter, as act is to potency.

all. If pressed *aether* offers no resistance; it cannot be affected by other bodies.[94]

The problem of *celestial* centripetal force (gravity) is that no medium has ever been detected or even suggested whereby the immense centrifugal forces required to overcome the innate tendency of a celestial body to rectilinear motion could be supplied. This is the reason gravity is treated as caused by a force of attraction innate in bodies. Having no answer to the dilemma, science has simply assumed that Newton's concerns are needless: causation *is* the same as calculation.

But the problem of the lack of a medium can be solved if it is understood that the contrariety in the way mundane and celestial circular motions occur corresponds to a fundamental difference in the natures of the relevant acting bodies. *Do follows be:* difference in *modus operandi* reflects a difference in *modus essendi*.[95] In bodies of ordinary matter centripetal force *precedes* circular motion; centripetal force must be secured before circular movement can be achieved. Before a wheel can turn spokes must be in place; before dancers can spin around a common axis hands must be interlocked. Though the two realities, centripetal force and circular motion, occur together in time, *ontologically*, i.e., in the order of reality, circular motion depends on centripetal force. The dynamic is *from within to what is without*.

But in *aether's* realm, outside mundane gravitational influences, the process is reversed: *the dynamic is from without to what is within*. Celestial circular motion does not depend on centripetal force; centripetal force (i.e., gravity) depends on circular motion! More than this: gravity is generated *as a consequence of* the circular motion *aether* produces in each celestial body.

This assessment finds confirmation in Aristotle's teaching that the agent that produces circular motion among the celestial bodies operates *at the circumference of* the circles of motion.[96] That is, their motion is initiated *not from the centre* of the action, as science claims *but at the periphery*.

---

94   A point well made by Christopher A Decaen in *Aristotle's Aether and Contemporary Science*, op. cit., Part I and footnote 40. This is also the reason why we cannot detect this substance, for every sense requires some reaction to the sense power in the thing sensed.

95   The nature of a thing determines how it operates.

96   In Book VIII of the *Physics*. Cf. St Thomas's Commentary *In VIII Physics*, L 23: 1168

I accept that what is asserted here is completely counter-intuitive; yet it provides a rational answer to innumerable questions about the cause of gravity.

iii. There are a number of objections that might reasonably be proposed to this thesis. First, there are celestial bodies that have very little actual motion—our Moon, for instance, rotates but once every month with but one side of it available for perusal from Earth. How can this thesis be maintained in respect of such bodies? The answer is that it does not seem to be necessary that there be *actual* motion of a celestial body; its *potency* to *aether's* influence is sufficient to generate gravitational force in it.

It might secondly be objected that the rotation of celestial bodies about an axis and the orbiting of satellites about a celestial body each occurs in one plane only whereas gravity operates in every possible plane about a celestial body's centre. The answer seems to be that *aether* is not constrained by the limitations of bodies of ordinary matter. Whereas *they* operate particularly and in one plane only, *aether* operates universally and in every plane. This is the only way to explain how *aether* produces spherical form; for a sphere, which is the mark of gravity, is simply a compound of every possible circle about a centre.

Aether's proper effect is circular motion—perfect motion; the form it produces in bodies is spherical form, perfect material form.

iv. The influence *aether* exercises falls within the metaphysical category *action* whose characteristics are set forth in chapter one under the sub-heading *The Categories of Being*. The reader might care to revisit that section before proceeding. As the commentator quoted there [John of St Thomas] remarks, it is one of the accidents that depends upon something extrinsic.

Action entails the production by some agent of an effect in another, called a *patient*, with movement from one to the other.[97] There are different species of *action*. That which concerns us here is called 'transitive'. An illustration: when something such as a spoon is placed in hot water the heat of the water (the agent) transfers into the spoon (the

---

97   This analysis is reproduced from the text of A. M. Woodbury Ph. D, S.T.D, of Sydney's Aquinas Academy, *General Natural Philosophy and Cosmology*, c. 22, 1, nn. 334 et seq.

patient); *action* in the water is *passion* (seventh accident) in the spoon. *Action* is really distinct from the movement involved as a line is distinct from its curve. *Action* adds to motion the respect 'from agent'. *Passion*, really distinct from the movement involved and adding to it the respect 'unto patient', is the reception of the effect. *Action* is an accident in the agent; *passion* another, separate, accident in the patient.

Now much as hot water produces heat in a spoon, *aether* produces the *passion* of circular motion in a celestial body and the consequent centripetal force of gravity.

v.   The experiment of Henry Cavendish in 1797-8 involving two sets of lead spheres of differing masses, 1.6 lb and 348 lb, each separately suspended some nine inches apart and on alternate sides, established that there is an apparent force of attraction, albeit infinitesimal, even among mundane bodies. The smaller spheres moved towards the larger causing the supporting arm to rotate: the twisting of the suspending wire enabled the force to be measured against the wire's torsion coefficient. Now if, as I claim, the force at work is not one of attraction between, but of extrinsic action by *aether* on, these bodies the experiment demonstrates that *aether's* action is not confined to the heavens but is universal.

Examples of *aether's* influence as extrinsic (efficient) cause may, I suggest, be seen in various earthly phenomena we ascribe to other causes. One is the soap bubble. When a pocket of air is captured by a soapy solution competing centripetal and centrifugal forces produce the evanescent miracle of spherical form. While the source of the centrifugal force is the trapped air, that of the centripetal force may be ascribed to the surface tension of the soapy water but in each case *only as material dispositions serving the bidding of* an efficient cause. There are other instances in nature. In the manufacture of shot quantities of molten metal forced through a sieve fall and solidify as tiny spheres before they are captured in a water bath; and, most common of all, water condensing in the atmosphere forms in tiny droplets, spheres, that fall as rain.

Cause and effect are always proportional. If spherical form in celestial bodies is the mark of *aether's* influence, why are these earthly instances not marks of it also?

vi. Let us recall the philosophers' teaching (III. iii above) that, while not a component, *aether* is an essential element of material being; that "it is involved in the composition of the whole universe as being part of it". The 'space' that science tells us makes up most of atomic and molecular structure of the elements and compounds of material bodies can, no more than that between celestial bodies, be 'nothing somehow existing'. According to our thesis *aether* is involved intimately in the structure of each celestial body: it cooperates with first metaphysical accident *quantity* in binding atomic and molecular structure. Since *aether* acts but cannot be acted on, the dilemma that confronted Le Sage's thesis does not arise: *aether's* extrinsic force bears not only on its surface but on the whole body, detectable common matter and 'empty space'.

This *passion* of inclination towards the centre of its mass of a celestial body resembles somewhat the *passion* of compression in a body submerged in the sea. Yet the analogy limps for, regardless of the depth and the intensity of its pressure, the creatures of the sea retain, under nature's edict, the rigour of their forms.

Some difficulties

i.   But if gravity is produced by an efficient cause *extrinsic* to a celestial body, why does it give the appearance of a force of *attraction*, i.e., of something *intrinsic*? If the *matter* (i.e., the subject) of gravity's centripetal force is the globe of the celestial body (and all its parts), the *form* (that which makes gravity be the immense accidental force that it is) is the inclination towards the centre of its mass. But gravity's strength or weakness, according to proven scientific principle, is a function of the *mass* of the heavenly body, something intrinsic to it. In breach of metaphysical principle, then, gravity seems to be determined by its *material* rather than by its *formal* cause.

The native motion of every element of common matter, as of the bodies they comprise, is straight (*rectilinear*) motion, as Newton makes clear.

> "A centripetal force is that by which bodies are drawn or impelled or in any way tend towards a point as to a centre. Of this sort is gravity by which bodies tend to the centre of the Earth; magnetism by which iron tends to the loadstone; and that force, whatever it is, by which the planets are continually drawn aside from the

rectilinear motions which otherwise they would pursue and made to revolve in curvilinear orbits."[98]

ii. Science treats circular motion as an application of Newton's Second Law. Shown below is an extract from a popular website, that of the University of California, Irvine, that applies his principles.[99]

Whenever an object moves in a circle with uniform velocity, it has an acceleration pointing toward the centre of the circle. This may seem confusing at first; we do not expect to encounter acceleration when the speed is constant. Remember that while the speed is constant, the direction of the velocity vector is continually changing, and it is because of this change in velocity that we have acceleration.

We know then, from Newton's second law (F = m a), that an object moving in a circle must have a net force on it which points in the same direction as the acceleration, i.e., toward the centre of the circle. The force associated with this centre-pointing acceleration is sometimes called the centripetal force. The centripetal force might be provided by a rope or by gravity or some other means; the designation 'centripetal' just means it is the net force that is associated with an object moving in a circle.

Combining Newton's second law and the equation for acceleration in terms of the speed around the circle, we have—

$$F = \frac{m \, v^2}{r}$$

The gravitational force 'of attraction' (Fg) between two bodies is calculated according to Newton's celebrated formula as follows—

---

98   Sir Isaac Newton, *Principia Mathematica*, Definition V; Axiom I. (Transl. from the Latin by Andrew Motte revised by Florian Cajori, University of California Press, 1934). My copy, a reprint in *The Great Books of the Western World*, vol. 34, for Encyclopaedia Britannica, Inc. 1952.

99   I have taken the liberty of rewriting the final formula to reflect its earlier mode of expression. See http://learn.uci.edu/oo/getOCWPage. php?course=OC0811004&lesson=006&topic=10&page=1

$$Fg \;=\; \frac{m_1\, m_2}{r^2} \times G$$

—where m_1 and m_2 are the masses of the relevant bodies, G is a fixed ratio called the gravitational constant, and r is the distance between the centres of mass. For the purposes of the present discussion I will not explore the subtleties elaborated by Einstein.

Velocity includes speed and direction. A change in direction is a change in velocity and since circular motion involves constant change in direction, science regards a circling body as accelerating towards the centre of its motion.

Gravitational force depends radically, then, on what science calls *mass*. But what is *mass*?

iii. Newton understood mass as convertible with quantity:

> "The quantity of matter is the measure of the same arising from its density and bulk conjointly. Thus air of double density in a double space is quadruple in quantity... This quantity I designate hereafter everywhere by the name of body or of mass..."[100]

Some say *mass* is constant proportion between force and acceleration, others constant proportion between weight and acceleration, a quantitative measure of an object's resistance to its change of speed. While weight varies from place to place, a body's *mass* remains unchanged— *pace* Einstein's theories. Another view has it that *mass* depends on the number of atoms a body contains. This is problematic because atoms are not uniform across the elements as the periodic table shows. A further view says mass is proportional to the volume a body occupies.[101] *Mass* is clearly not volume for volume is variable under the influence of pressure and temperature. It is not weight for weight is an effect of gravity and varies with altitude, i.e., distance from the centre of the Earth (or other celestial body).

Perhaps the most objective assessment, at least for the purposes of Newton's Laws, is that it is a measure of the force necessary to deflect a body from rectilinear motion, the straight motion natural to it; in other

---

100 *Principia Mathematica*, op. cit., Definition I
101 Cf. The entry on *Mass* on the Wikipedia website, http://en.wikipedia.org/wiki/Mass

words, it is a measure of its inertia. While the definition of each of these categories seems constrained by another in a bemusing circularity, it should be observed that science now expresses, in the unit 'the *newton*', a measure of a body's lineal inertia, the force necessary to cause it to accelerate at a given rate.[102]

iv. *Aether* operates, according to the thesis proposed here, universally and uniformly (via its proper accident of *action*) *in* and *about* each celestial body to produce in it the *passion* of circular motion. *In the body* it produces rotation on one of the infinite number of possible axes about its centre or focus and thereby generates internal centripetal force, gravity. The larger the body the greater the gravitational force generated and—it would seem—the more perfectly the body approaches spherical form. The inclination to the centre or focus, taken ontologically, is the formal principle of rotation of the celestial body, both actual and potential. But no celestial body exists alone in the *aethereal* sea. *About* the body, then, *aether* produces the *passion* of circular motion relative to each neighbouring body with an intensity that reduces in proportion to the square of the distance between their masses.

v. Consistent with Newton's principle, if a body of common or ordinary matter is moved circularly a consistent force must be applied to it by the relevant agent; and the greater the mass, the greater the force the agent must exercise to overcome its rectilinear inertia. In other

---

102 The inability of science to plumb the nature of mass is understandable for science is not concerned with the natures of things. A *substance*, as we have noted, is something which exists in itself (*be-in-self*) not in another; an *accident* is a reality which exists not in itself, only in some substance (*be-in-other*). The first accident is *quantity*. Metaphysically understood, then, the *mass* of a body consists in corporeal *substance* as affected by *quantity*. But there is, as Newton remarked, another influence too, that *quality* which is the substance's proper density. *Substance* explains the specific differences between masses, *quantity* explains the individual differences between them, while density, the *quality* proper to each substance, explains why one type of substance differs from another in specific gravity. "The action of a generant does not stop at the bare substance but produces it equipped with the accidents upon which the substance depends, that it may exist and operate." (John of St Thomas; *Curs. Phil. II*, ed. Reiser, p. 268b, quoted in A M Woodbury, *General Natural Philosophy and Cosmology*, op. cit., nn. 127 and 344.) The *substance* of copper (that immaterial reality which is copper), for instance, differs from the *substance* of water (that immaterial reality which is water), as the density which is the *quality* proper to copper differs from the density proper to water. One mass of copper differs from another, as one mass of water differs from another, through their respective *quantities*. [This analysis from A M Woodbury Ph D, S.T.D., *General Natural Philosophy and Cosmology*, op. cit., nn. 238 to 245. Cf. https://austinwoodbury.com/pg-search.php]

words, what is *intrinsic* to it—the body's mass—specifies the extent of the force to be applied by the *extrinsic* agent. If it is accepted that in *aether's* realm circular motion is prior to, and causative of, centripetal force, then the force *aether* must exercise to 'draw aside from the rectilinear motion which otherwise [the celestial body] would pursue' will be greater according as its mass is greater; and the greater will be the centripetal (gravitational) force generated in that body as a result.

It is understandable that, unless he is informed by metaphysical principle with its demand for an adequate (*extrinsic*) efficient cause of these effects, an observer will be moved to attribute *aether's* action (which he cannot see) to the body that he can see, and treat the force as if it is something inherent in matter.

Einstein embraced the materialistic paradigm as he immersed himself in the thought of Hume and Mach. He accepted the materialist conclusion flowing from the Michelson-Morley experiment that no ether existed. Space seemed, from observations, to exercise a certain causative faculty. Uninhibited as Newton had been by a residual metaphysics, he saw no difficulty in ascribing such causality to something bereft of any objective reality. Gravity was a natural outcome, he said, of the presence of the mass of a body in space. It 'warped' the space around it, impelling other bodies, should they approach too close, to depart from their rectilinear paths. The greater a body's mass, the more it 'warped' the space around it. Gravity was not a force propagated between bodies but the inevitable effect of the interplay of their mass and the surrounding space.

vi. Like light, gravitational force is an accident rooted in *aether's* influence. It is not to be wondered at then that (as with light) gravitational force appears to operate at *c*, the 'speed of light' as Einstein has established in his *General Theory of Relativity*.[103] However, it must be insisted that *c* is not a property of light but of its proper substance, *aether*. It is the speed at which *aether* determines light's propagation, as

---

103 Einstein's view is, thus, to be preferred to that of Newton who held that gravitational force was instantaneous. Newton seemed to treat 'space' as if it was an ethereal body while Einstein, at least until he amended his view in 1920, treated it as non-being-somehow-existing. His amended view did not, however, seem to regard 'ether' as more than an accident of 'space', albeit he was correct when he said that it was not to be considered as comprised of parts track-able through time, or of ponderable matter.

it is the speed at which *aether* determines the operation of all gravitational force.

The Problem of the Tides

i. But is this thesis not contradicted by what we observe of the influence of the Moon, and to a lesser extent the Sun, on the seas which cover some seventy per cent of Earth's surface? If gravity is caused not by something intrinsic to a celestial body but by this extrinsic influence, *aether*, how explain the clear influence of the Moon and the Sun on the regular movements and alterations in movement of the tides?

ii. The Moon's involvement in the tides may be seen in the way the diurnal period between successive tides reflects the lunar day, about 24 hours 50 minutes. A cause *exercises influence unto the 'be' (esse; existence) of a thing dependent in regard to be.*[104] Causes may be distinguished according as they are essential or not essential to the effect; that is, a cause may exercise its causality *per se* or *per accidens*.[105] Of any effect there are four *per se* causes, no more and no less, as outlined in Chapter 1.

Causes *per accidens* fall into three categories, *condition, occasion* and *chance*. Of these one only concerns us here, *condition (removens prohibens)*, without whose operation a cause *per se* cannot act.[106] Now the Moon is more than a *condition* of the motions of the waters that cover the Earth; it is essential to their regularity. It must, then, be a cause *per se*. It is not a *formal* or *material* cause for these are always *intrinsic* while the Moon is *extrinsic* to the Earth and its motions. It is impossible that it be the *final* cause, their end or reason, for this is something intended by nature's Author. It remains that the Moon is an *efficient* cause albeit subsidiary to the *principal* efficient cause which is the intellectual substance the philosophers have identified. Hence the

---

104 This principle is elaborated by St Thomas Aquinas in *Summa Theologiae* I, q. 104, art. 1.

105 Aristotle *Metaphysics* Bk. V, ii. and *Physics* Bk. II, vii (198a 5 et seq,); St Thomas *In II Physics* L. 10.

106 For a detailed analysis see the material under the heading 'The Mode of Aether's Involvement' in *Science and Aristotle's Aether* at http://www.superflumina.org/PDF_files/aether_science.pdf

Moon operates in the capacity of instrument. But there are degrees of instrumentality.

iii. Science's explanation that the tides are caused by the 'pull' of the Moon is problematic; the thesis of attraction is grounded in the confusion of causation with calculation but it suits the materialist inclinations of modern science which relegate to insignificance, or deny completely, extrinsic causation of any effect in nature. A 'bulge' of waters, a high tide, occurs with the Moon's passing, though not necessarily opposite the Moon's meridian, but the waters mass uniformly and as uniformly diminish with the Earth's rotation, albeit with amplitudes which differ from place to place. But, even stranger, the massing that occurs in the hemisphere adjacent to the Moon is balanced by a corresponding massing in the opposite hemisphere. The total effect is a relatively even pulsation, analogous to an animal's breathing, with nodes on opposite sides which process unceasingly about the globe.

iv. Although we take the rotations of Earth and Moon as simple circles about their axes, their motions are more complex. The planet and its satellite each have an effect on the other, a function of their respective masses. The Moon is 1/81st the mass of the Earth, its relative density 3.36 to the earth's 5.5.[107] Its average distance from the Earth is some 384,000 km. In accordance with Newton's laws it influences the Earth in direct proportion to its mass and inverse proportion to the square of the distance between the centres of their masses. The Earth influences the Moon in a similar fashion.

The combined masses, separated though they be by some 380,000 odd kilometres, circle about a focus (the centre of the two masses, their combined centre of gravity) called the *barycentre*. This is located within the body of the planet at a point opposite the Moon an average 4,670 km from the Earth's geometric centre (some 1,700 km beneath its surface).[108] The Moon moves in the same direction as the Earth rotates,

107 For this and what follows see for example Steve Massey, *Exploring the Moon*, Sydney, 2004.

108 Cf. http://en.wikipedia.org/wiki/Barycentric_coordinates_(astronomy) The reader should study the graphics on this website for an appreciation of the crucial part played by the barycentre in the operations of the heavenly bodies.

anti-clockwise.[109] But in the time it takes the Earth to rotate 360° the Moon progresses only 12.2°.[110]

The consequence of this disparity is that the *barycentre* beneath the Earth's surface moves, relative to that surface, in *the opposite direction to* the planet's rotation somewhat after the fashion of the phenomenon known as mechanical precession.[111] It moves slightly slower in that opposite direction for its locus, aligned between Earth's centre and that of the Moon, traverses 348° in the time the Earth rotates 360°.[112] The accompanying Diagram A illustrates.

Diagram A

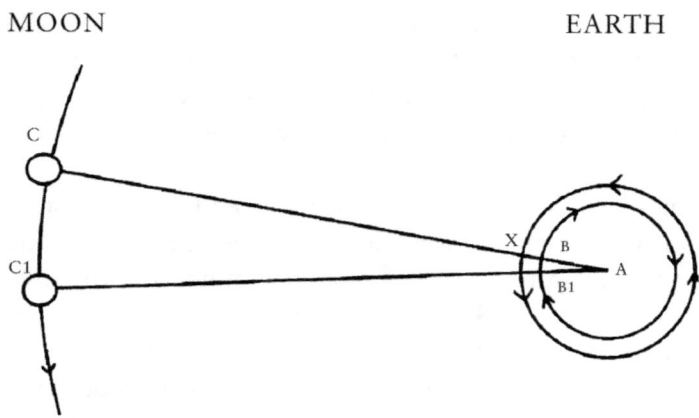

A = Earth's c/g B = barycentre C = Moon's c/g

The larger circle represents the globe of the Earth taken from a point above the North Pole. The smaller is the Moon, C, which moves after 24 hours to C¹. The planet rotates in an anti-clockwise direction; the

109 So, too, the motion of the Earth around the Sun is counter clockwise viewed from the north pole.

110 Hence the Moon advances from west to east some 49 minutes every day. Taken with respect to distant stars, the Moon takes 27.32 days to orbit the Earth (sidereal month). But because the Earth is itself moving circularly around the Sun and, in one cycle of the Moon, traverses about 1/12th of its annual cycle, the Moon takes about 29.53 days (synodic month) to pass from new Moon to new Moon.

111 *Mechanical precession* is the movement of a round part in a round hole where the direction of rotation of the inner part is opposite to the direction of rotation of the radial force.

112 Cf. the title 'Tides' at http://en.wikipedia.org/wiki/Tides

Moon does the same. A is the Earth's centre of gravity (CG). B is the Earth-Moon centre of gravity, the *barycentre*. The smaller inner concentric circle represents the locus of the passage of the *barycentre* beneath the Earth's surface.

The *barycentre* is the focus of a centripetal (gravitational) force *additional to* that exercised by Earth's centre of gravity which produces the tides in the seas and in other bodies of water on the Earth's surface.[113] After the passage of 24 hours, when a point on Earth X, moves back to X and the Moon moves from C to C¹, the *barycentre* arrives at B¹.

v. In Diagram B the two circles around A are enlarged with the directions of their respective movements shown. For the purposes of illustration the position of the Moon relative to the Earth and the movement of the *barycentre* are taken as fixed and only the globe of the Earth assumed to be moving. The force exercised on the waters of the seas around the surface of the Earth is a repulsive centripetal force directed towards the Earth-Moon *barycentre* B.

Diagram B

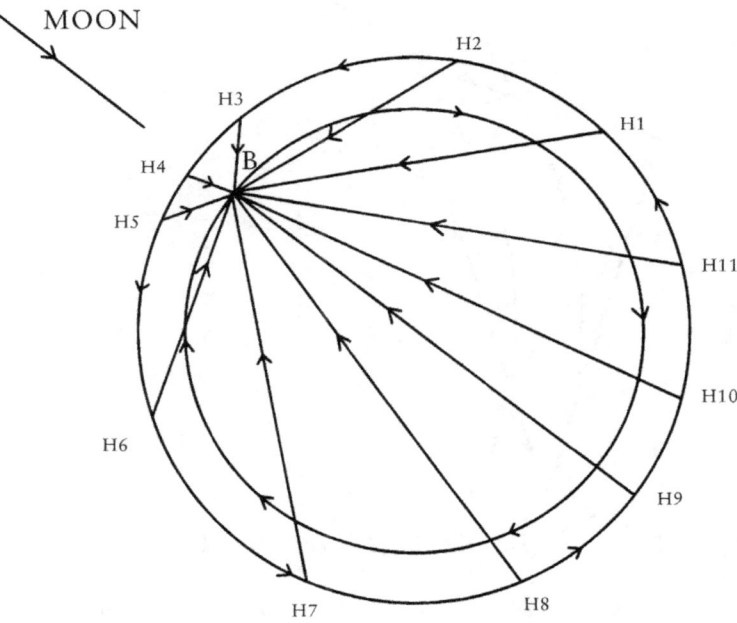

113 For it is a function of the two masses rather than that of Earth alone.

This force is illustrated as exerted on a buoy floating in the seas on the equator (reflecting the water in which it floats). The constantly altering magnitude of this force is represented by the varying lengths of the lines throughout the course of 24 hours in 12 different phases beginning and ending at H1 and focused on the *barycentre* B. The inverse square law applies (the intensity of a force at a point is inversely proportional to the square of the distance of the focus of the force from that point[114]) so that the shorter the line the stronger is the centripetal force indicated, and vice versa. The force on the buoy and the waters of the sea in which it floats first increases from H1 to H4; then, from H4 to H9 it decreases progressively before, beginning at H10, it slowly increases until it arrives again at H1 and the process is repeated. Diagram C below illustrates the forces at work.

Diagram C

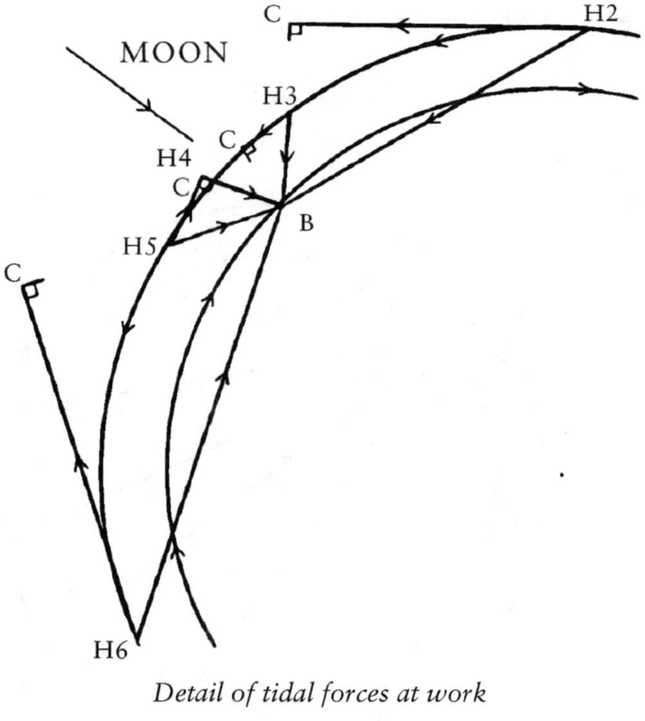

Detail of tidal forces at work

114 The law is normally expressed relatively to the source of some energy source as, e.g., the Sun in respect of light, a fire source in respect of the heat it emits. Since metaphysical theory treats the centre of gravity of a heavenly body not as the source of that force but as its focus, the expression of the law has been adjusted accordingly.

The extrinsic power generating gravitational force may, consistent with Newton's thinking, be treated as a force of attraction (increasing with proximity) focused on the Earth-Moon barycentre B (here treated as stationary though moving in opposition to the rotation of the planet). This resolves into two forces, one of which, the *tidal force*, acts at right angles to the surface of the planet towards the point C at each of H2 to H6, the length of each arm indicating the proportion of the gravitational force involved.

At H2 the *tidal force* is accelerating the surface waters' flow towards a point above B. At H3 though the barycentripetal force has increased (its distance has shortened substantially) the *tidal force* is greatly diminished. At H4 it is nil. At H5 the barycentripetal force has again increased but now, because of B's recession from the surface waters in question the horizontal component, the *tidal force*, works to decelerate their momentum and the water level abates.

The waters of the sea are moved according to the laws of fluid dynamics which take account of momentum and hydrostatic elasticity under the influence of external forces the chief of which are gravitational. The lines of force indicated in H1 to H3 in the diagram resolve into a force on the buoy (and the waters in which it floats) directed towards a point on the surface of the globe above B. It is this *horizontal component* of the *barycentripetal* force which causes the seas to swell to a high tide with a momentum that carries beyond B and reaches, perhaps, H5. From H4 the horizontal component force decreases progressively. Under the constantly moving influence of the *barycentre* the tides progress from west to east, though the interference of continents, the varying depths of the oceans, the *coriolis* effect and other factors produce a complex of movements. These are manifested in a pattern of co-tidal lines (lines joining points of identical tidal phase) radiating around centres which turn clockwise in the southern hemisphere and anticlockwise in the northern.[115]

This contrariety in motions—of the Earth's surface from west to east, of the *barycentre* from east to west at a slightly slower rate—produces a contrariety of forces which explains (i) why the sea wells and dissipates at a uniform rate with the passage of the *barycentre*, the momentum of

115 Cf. http://en.wikipedia.org/wiki/Tides for the reproduction of these lines on a homolographic projection of the Earth analogous to lines on a topographic map [M2_tidal_constituent.jpg] .

the waters carried to the east matched by the march of the tidal node to the west, and (ii) why the tides advance by 50 minutes or so every 24 hours.

There is another feature. The Moon moves progressively from *apogee* (furthest away) to *perigee* (closest) every 7½ cycles. As the Moon moves closer the *barycentre* moves closer to Earth's surface. The lines of force from H2, H3, H4 and H5 are shortened, indicating an increase in the *barycentripetal* gravitational force. The horizontal component of this force assumes greater magnitude. It is this greater force, not a stronger 'pull' of the Moon that explains why the tides at such times increase in amplitude.

vi.   To the high tides induced by the ever-moving *barycentre* there correspond highs on the opposite side of the planet. Let us call the first set *primary* tides and the corresponding ones *secondary*. There is currently no satisfactory scientific explanation for these *secondary* tides. All those proposed are premised on the thesis that the Moon exercises a 'pull' on the Earth. But no such hypothetical 'pull' on one side of the globe could account for the massing of the seas in the opposite hemisphere *away from* the alleged pull. The best explanation, and one that accords with the *barycentripetal* theory, is that the secondary tides are reactive movements of the waters of the globe (in accordance with the laws of fluid dynamics) to the intensity of the *barycentre's* influence in producing the primary high. The pulsing of the tides reflects, too, the wave motion characteristic of fluids.

An analogy can be drawn with the operation of a single cylinder four stroke engine with its four phases—intake, compression, power and exhaust. Such an engine relies on the momentum generated in the power stroke to carry it through the other three phases. In similar fashion the energy generated in the massing of the primary high tide of the Earth's seas and lakes under the effects of the contra-rotating influence of the Earth and the Earth-Moon centre of gravity every twenty four hours carries its momentum through the motions of secondary high and secondary low tides. Every example limps and the analogy is not perfect, yet the operation might be illustrated as follows:

Engine	Tides
intake stroke	secondary high to secondary low
compression stroke	secondary low to primary high
power stroke	primary high to primary low
exhaust stroke	primary low to secondary high

As with *primary* tides the *secondary* tides have four causes. We need concern ourselves only with the *extrinsic* ones (*efficient* and *final*). Their *final* cause is clear. Without them the centre of mass of the planet would move and the resultant instability would adversely affect the regularity of its rotation and its relationship to Moon and Sun.

Again, the principal *efficient* cause is clear, it is the intellectual substance that ensures the planet rotates with due order to ensure the welfare of its parts.

vii.   The Moon is, thus, a subsidiary instrumental *efficient* cause of the movement of the tides in the following subordination:

Primary	*aether*
Secondary	the Earth-Moon *barycentre*
Tertiary	the Moon according as *aether* constrains it to circular motion around the earth generating a centripetal force proportional to its mass which, with the Earth's mass, produces the moving *barycentre* which affects its surface waters.[116]

viii. How does the Sun affect the tides? Again there is no 'pull' exercised, although the analogy of attraction is closer than with the Moon's involvement because the location of the Sun-Earth *barycentre* is within the body of the Sun and close to its centre. The influence operating

---

116 In fact the Moon affects Earth's progress such that, as they circle the Sun, both rotate about their common centre of gravity.

on the Earth's seas is the centripetal force upon them focussed on the Sun-Earth *barycentre* generated by *aether's* causation of circular motion of the planet and the Sun about that centre.

The rhythm of the unremitting cycle—primary high, primary low; secondary high, secondary low—is reinforced twice each *lunar month* (i) at the beginning when the Moon is in the same quadrant as the Sun (new Moon), and (ii) mid-month when it is in the opposite quadrant (full Moon).

ix. An interesting problem arises over the relative influence of Sun and Moon. Notwithstanding its distance, some 149.6 million kilometres (93 million miles; 8.32 light minutes) from the Earth, the Sun's influence, through its mass, is some 179 times that of the Moon. Yet the Sun's observed influence on the tides is less than half that of the Moon. In an endeavour to solve the problem, the current view of science—grounded, of course, on the thesis that the centripetal force of gravity is one of *attraction*—is that the tides on one celestial body are influenced by another not according to the square but *the cube* of the distance from that other body, a thesis which departs from scientific principle.[117]

But if the tides are influenced not by the distances between the relevant bodies but by the distance of the seas from their respective *barycentres*, this gratuitous modification of the inverse square law is unnecessary. For in respect of the tides the Moon is not competing with the Sun as—

$$\frac{\text{mass of Moon}}{\text{distance}^3} \quad \textit{is to} \quad \frac{\text{mass of Sun}}{\text{distance}^3}$$

but as—

$$\frac{\text{Earth-Moon}\ \textit{barycentripetal force}}{} \quad \textit{is to} \quad \frac{\text{Sun-Earth}\ \textit{barycentripetal force}}{}$$

---

117 Cf. https://en.wikipedia.org/wiki/Tide under the heading 'Forces'

$$\frac{\text{distance of proximate}}{\text{seas from Earth-Moon}}\quad\frac{\text{distance of}}{\text{proximate seas from}}$$
$$\frac{}{barycentre^{2}}\qquad\frac{}{\text{Sun-Earth }barycentre^{2}}$$

To put the issue as Newton has suggested, the 'pull' of the Sun is not competing with a 'pull' of the Moon but with that of the much closer Earth-Moon barycentre which involves not only the mass of the Moon but also that of the Earth.

x. As with the Moon then, the Sun is an instrumental, but subordinate, *efficient* cause of the movement of the tides. The subordination operates here as follows:

Primary	*aether*
Secondary	the Sun-Earth *barycentre*
Tertiary	the Sun according as aether constrains the Earth to circular motion around it generating a centripetal force proportional to the joint masses of Sun and Earth which produces a barycentre of great force but remote from the proximate seas on the surface of the Earth.

The Conclusions Summarised

i. Gravity's *final* cause is the ordination and subordination, for the good of the whole, of the material substances that constitute the globe and, in the case of Earth, its inhabitants.

Gravity's *formal* cause, operating to give effect to the final cause, is *the inclination towards* that centre or focus, the consequence of the circular motion induced in a celestial body and its component parts by *aether*, the heavenly body. Its formal cause explains why gravity appears to be a force of attraction. As the house-plan realised is the term of the work of construction of the house (its *final* cause) unless something (e.g., an obstruction; defective materials) impedes it, so attainment of the centre of mass of the celestial body is *the term* of the work of gravity, unless something (other matter) impedes it.

Gravity's *material* cause is the celestial body and its component parts moved circularly, i.e., against their natural, rectilinear, inclination.

Gravity's *principal efficient* cause is the intellectual substance which orders the universe. Gravity's *instrumental efficient* cause is the *aether*.

ii. It is characteristic of natural things that their speed of progression increases as they approach their term. This is the philosophical reason why a body accelerates as it approaches the centre of a celestial body. It is this principle, too, working with the relative densities of the component elements of the celestial body, which assists the right ordering of its parts. For, were it otherwise, gases as the least dense of its components would not rise above all others; and water, less dense than the generality of minerals, would not rise above them but be admixed with them in confusion. Hence the *formal* cause of gravity, that which determines the matter so that the end (*final* cause) of the operation is achieved, ensures that the inclination towards the centre is greater the closer another body approaches it.[118]

iii. In Chapter 3, the chapter on light, I remarked the relationship between *aether* and light and offered the conclusion that *aether* is universally the vehicle of light's transmission. In the performance of its function as *lucifer* (lit. 'bearer of light'), as in those functions that relate to the very structure of bodies and the conduct of celestial bodies, each of them essential to the works of creation, *aether* might be called a *pure instrument*.[119] Consistent with this thesis, whether at the level of *the infinitesimally small,* where it assists in binding atomic and molecular structure, or at the level of *the almost infinitely great,* where it holds together each solar system, each celestial body, the very universe itself, *aether* operates unobtrusively and in undetectable fashion.

* * *

118 Which facility is recognised in the inverse square law.

119 There is a parallel in philosophical psychology in the field of knowledge where the objective concept, which represents the form of the thing known, has no reality save as instrument to serve the intellect. It is a pure sign or a pure instrument.

APPENDIX

An Experiment

Here is a suggested experiment, to test the thesis advanced here, beyond the means or the facilities available to the author.

Fashion a hollow, clear plastic, discus. A central pin or axis may be inserted but it is unnecessary. Make the whole sufficiently strong to cope with changes in pressure. Import some fluid into the hollow body, pure water, alcohol, or oil, and seal it at standard atmospheric pressure so that it is, perhaps, 1/8th filled. In the space-station where a similar ambient pressure obtains, in a condition of weightlessness, have one of the crew agitate the discus so that the liquid is dispersed throughout its internal volume then, using both hands, have him spin the discus around its central axis and allow it to hang spinning in the cabin space. Observe what happens to the fluid.

Prediction: the fluid will migrate to the axis and collect in a sphere.

5. THE UNIVERSE AND OUR PLACE IN IT

Further Lights on Gravity

Our age has the benefit of phenomena that Sir Isaac Newton and Albert Einstein would have given anything to obtain, data showing the behaviour of bodies removed from the influence of Earth's gravity. Observations in the International Space Station of what befalls fluids removed from gravity's influence are revealing. A quantity of water takes the form of a sphere. One may view a video clip of this phenomenon on the internet. Someone viewing the video might reasonably ask himself why the scientist-astronaut experimenting on the sphere of water did not remark the way the tiny globe emulated in miniature the globe beneath the space-station on which he was a passenger, the greater part of whose surface is, similarly, comprised of water. Or wonder why the experimenter did not ask the question, if no more than rhetorically, whether it was possible the cause of sphericity in the one might be the cause of it in the other also.

Certainly, Newton would have seen in the phenomenon support for his thesis that, while the effects of gravity may be calculated as if they entail an intrinsic force, their cause must be some extrinsic agent. The phenomenon of sphericity is repeated in other instances. The vaporised portion of water brought to the boil in a container, without the influence of convection or buoyancy (each of which depends on gravity), remains adjacent to the heating surface as may be seen in photos available on the internet. Other emollient material also tends to mimic the form of the celestial bodies. A candle flame becomes as near as possible a globe of fire.

Scientists explain the instances of sphericity of fluids in zero gravity as they explain similar phenomena occurring on the surface of the Earth as attributable to surface tension. While it is true that this facility in each of the fluids mentioned *disposes it* to take on spherical form, more is required than mere disposition. A house does not build itself just because there is a disposition in the materials lying around the building site to be formed into a house. Certainly, the house could never be built if that disposition did not first exist—you cannot build a house from materials indisposed to the task, like oil, water or air. But more is required. A little boy in his toy car calling on his father to push him does not move himself just because he (and his car) are *disposed to be* moved. Without the builder building, the house is not built. Without the father pushing, the child and toy car are not moved. In each case an *extrinsic* cause—something outside the subject—is essential if the effect is to be achieved.

In the same way none of the fluid subjects mentioned above can take on spherical form unless an efficient cause, a cause extrinsic to them, impresses this form in their matter.

There is another issue. Cause and effect are proportionate; the more universal the effect, the more universal the cause. The warmth I enjoy when I enter my house is from the fire in the grate. This particular effect testifies to a particular cause. The warmth I encounter when I go outside on a clear day, however, is something I share with the whole world because it is from the Sun which heats the planet and the whole solar system. The more universal the effect, the more universal the cause. Now sphericity of form is found throughout the universe. Why is it not reasonable to conclude that the cause that induced sphericity in the ball of water on which the scientist was experimenting is identical with that which induced global form in our planet and in every other one of the celestial bodies?

As remarked in Chapter 1, it is the failure in this *moral* issue— 'moral' in the sense that it is rational to acknowledge the influence of an adequate cause in nature's operations—which blinds the present genera- tion to the causes of gravitational force.

Gravity is an accident, a *quality*. It does not exist in itself, only in some substance. Newton's assessment of how gravity works remains the practical model notwithstanding that it has been superseded by Einstein's theories. His assessment is grounded, reasonably, in the three laws of motion and particularly in the second, expressed in the formula $F = m\,a$—force is the product of mass and acceleration. But mass, too, is an accident, a *quality,* closely allied with the *quantity* of a substance.

Matter, we recall from the principles set out in Chapter 1, is incapa- ble of determining. Matter's office is *be determined* and it is the *matter* of the planet and of each of its component parts—material substances such as the author (!)—which *is determined by* gravity's accidental form. Now, if the planet (and its components) is gravity's *material* cause

111

it is impossible that, together, it could be gravity's *efficient* cause. More-over, an efficient cause is always *extrinsic*, i.e., outside the effect.[120] The builder constructing a house is *extrinsic* to the form and the materials he uses in his building. The father pushing his little son in the toy car is *extrinsic* to the motion he induces in them. Metaphysical principle demands that gravity's *efficient* cause, too, is something *extrinsic* to the planet and its components. This demand resonates with four of the objections (nos. 1, 5, 6 and 7) to the premises underlying current gravi-tational theory set out in part I of the previous chapter.

Though they understood its effect, heaviness, which they character-ised as a *quality* attaching to things, the philosophers Aristotle and St Thomas Aquinas had no notion of gravity as centripetal force associ-ated with a celestial body. Let us note the contrariety, the opposition, in the approaches of metaphysics and modern science. Science addresses gravity first and treats circular motion and the sphericity found in celestial bodies as little more than incidents. The two philosophers, in contrast, consider circular motion as primary and essential, the signal characteristic of the heavenly substance, through which it exercises its causality. Their cosmology was insufficient for them to understand heaviness as due to gravity, or to understand in what gravity consists (centripetal force associated with the globe of the planet). They were unable, accordingly, to explore the causal connection between the circu-lar motion which was *aether's* proper effect and gravity. Had they been possessed of these knowledges, I suggest, they would have endorsed the arguments advanced here.

Aether's Two Functions

From the discoveries of modern science we might reasonably induce that *aether*, the heavenly substance, exercises two offices or functions. First, it is the instrumental efficient cause of the order in the universe and its component parts (*ordinator*) achieved by its generation of circu-lar motion, centripetal force and sphericity of form. Second, it is the means whereby light and other electromagnetic energy is conveyed

---

120 Except in the case of living things which move themselves; and then one part, the soul, moves another.

(*lucifer*). For convenience's sake I repeat here what I assert to be *aether's* properties:

- it is transparent by essence;

- it determines *c*, 'the speed of light' (299,792,458 km/s), the speed at which light and other electromagnetic energy is propagated, and the speed at which gravitational force is propagated. Moreover, since gravity is (in the metaphysical theory I have advanced) centripetal force that follows upon circular motion, it seems reasonable to conclude also that it is the speed according to which the *aether* acts to invest a body with circular motion and, in the case of emollient materials, with sphericity of form;

- it establishes, through *c*, the ground in which *time*, the measure of change, is established;

- it is convertible with what we call 'space';

- it is the catalyst, with its proper *quantity* (primarily) and its proper *qualities* (consequently), in the constitution of each material substance.[121]

Einstein regarded *c* as the one constant in the Universe. But *c* is not a *substance*, it is an accident. Had he understood its reality as Aristotle expounded it, rather than in the stilted version materialism was prepared to accept and was quick to deny after the Michelson-Morley experiment, Einstein might have acknowledged *aether*, the heavenly body, as the one constant in the Universe. He might have acknowledged the heavenly body as the cause of gravity too, rather than attributing it to a subsistent nothing, empty space. All this, however, would have involved a sea change in the philosophy to which he had pledged his adherence. His recanting of a denial of the existence of an 'ether' later tends to support this view. One can only wonder at what might have been.

Einstein didn't pretend that his conceptualisations could be accommodated to human experience and they are, in truth, no more than mathematicians' fantasies, graphs come to life to represent an ersatz

---

121 'Primarily' and 'consequently' here signify the ontological, not the temporal, order.

reality whose only justification is the accuracy of the behaviour they predict.

The phenomenon known to astronomers as "lensing" where a ray of light from a distant source is refracted around an intermediate celestial body, demonstrates, I suggest, a subordination in these two functions, of the light-bearing function in the *aether* as subsidiary to its ordering function. The phenomenon is demonstrated in the constellation *Pegasus* by the cluster known as *Einstein's Cross* where light from a distant quasar is refracted fourfold around an intermediate constellation.

Light does not travel directly to us (the observer) from the quasar in *Pegasus* but indirectly along a path which (in each of the four instances) represents two sides of a triangle, with the direct route (were it not impeded) being the base. Though the speed of light's transmission is not altered, each ray takes longer to reach us than it would because of its diversion by a force which science attributes to gravity but which, on our thesis following Aristotle, is the primary force *aether* exercises in inducing circular motion (rotation) in the intermediate star or constellation. There is a lag in the time the light might otherwise have taken to reach us as a consequence of this ordering function. Other instances of 'lensing' provide more dramatic evidence of this subordination, notably the appearance of a *supernova* behind a star or constellation where a second appearance may not occur until years after the first, the time lapse indicating that the paths taken by the two light rays differ vastly in length.

The theory behind 'black holes' may also be comprehended by this thesis. If the force the *aether* exercises as *ordinator* in constraining a celestial body to circular motion is sufficiently great this may circumscribe completely its function as *lucifer* and so preclude any escape of light from the vicinity of the subject body. A search of the internet will show the 'black hole' known as Messier 87 in the constellation *Virgo*.

'Black holes' pose an interesting question in line with Aristotle's approach of placing the generation of circular motion as the *aether's* primary effect. If we accept Newton's Second Law, expressed in the formula $F = m\,a$ as of universal application, the force exerted by the *aethereal* matrix on the body within a 'black hole', so great that it prevents any escape of light, may be more a function of a than m; that is, more a function of the rotating force exercised on it by the *aether* than its mass.

The Heavenly Substance Is Motionless

Nothing moves without being moved by another. Nor can a series of moved movers proceed to infinity as Aristotle shows (*Metaphysics* Bk. II, c. 2, 994: & see St Thomas *In II Metaphysics*, Lesson 2). There must be a first mover which is itself unmoved, and this is God. It seems reasonable to contend that the heavenly body, *aether*, is the first instrument of God's agency. Through it all others are governed and sustained. As a matter of principle it seems fitting then that *aether*, His first instrument, should also be motionless. Its operation (*modus operandi*) differs radically from that of bodies of common material being as its nature (*modus essendi*) differs radically from theirs. Its behaviour excels their behaviour as its nature excels theirs.

A body deep in space, whether natural or artificial (like a space-probe), with no other body in proximity to it, is motionless—and this no matter how fast it may be said to be moving relative to the Sun or to our own planet. The body is surrounded and sustained by the *aethereal* matrix much as a sea-creature is surrounded and sustained by the sea. Yet the analogy limps, for the sea-creature *moves through* the sea that contains it. But the body 'in space' does not move through the *aethereal* matrix. It is motionless in *aether*. And, reciprocally, *aether* is motionless with respect to the body. This assertion accords with a view maintained by Dutch physicist Hendrik Lorentz (1853-1928) against Michelson and Morley's 'disproof' of the existence of ether.

This understanding may seem counter-intuitive for us surrounded as we are by bodies forever in motion, but the possibility is supported by St Thomas's view mentioned above that *aether's* accidents (manifest in the phenomena that attend it) are wholly disproportionate to those with which we are familiar. [*In II De Caelo* l. iv, n. 3]

Acceptance of *aether's* immobility assists in understanding other phenomena. It explains why *aether's* proper accident, the *quality* light, is immutable in its speed of propagation. Its proper *substance* is immutable! It provides a reason why light's speed of propagation is the same in every direction and setting: *aether* is *motionless in every setting*. Hence, no matter how fast these bodies may be moving relative to each other, light and each of the species of electromagnetic energy emitted by either will be propagated, and will be received, at a speed which is invariable.

Aether is the universal agent. Its accidents are propagated simultaneously in every dimension and in every plane. The imponderability of these accidents—they have no mass as they have no independent existence—permits them to be propagated at the speed which is *aether* prerogative, 299,792,458 metres per second. In a sense, save by the inertia that attaches to every material being *because it is material*, they are unlimited. Light, as remarked by Christian Huygens, travels from its source untrammeled in *every* direction. In contrast, a body of common material being is limited by its nature to motion in one direction at a time. Since such a body is a *substance,* dependent and determined by *aether*, it is bound by that first body's determinations. Here is the reason for the difficulties experienced by scientists in attempting to accelerate a particle, i.e., a material *substance* at its smallest, to the 'speed of light'. It can't be done! Light-speed is only available to accidents of *aether*. No substance, no material body, can breach its limitations.

The Proportion between Man and Reality

There is order between the known and the knower. That is, there is proportion between the world of reality and man whose knowledge is not just of singular things, like that of the brute animal, comprehends *the very nature of* the things that fall under his senses. The issue is encapsulated in St Thomas's passing comment in the *De Veritate,* (I, 2) *res inter duos intellectus constitu[itur]*..., "the [natural] thing [is] established between two intellects".

Man was created by God as the highest of His material creatures to live and move and have his being in the world He created. There is a proportion between creation and the creature; between reality and man the knower; between Him who created the thing (*res*) and the intellect He created to know it. Put another way, what man knows is what is: *reality*. In 1913 the Dutch mathematician, physicist and astronomer, Willem De Sitter (1872-1934), unwittingly provided testimony of the Providence that established *aether* as the ground of the constancy of *c*, 'the speed of light', by demonstrating that if the speed at which light was propagated varied with the motion of the body emitting it, man could never know the truth of the behaviour of double stars in different phases of their orbital paths. There is a reproduction on the internet of a binary star in the constellation *Gemini* which provides an illustration.

In his old age, when he had outgrown many of the materialist and subjectivist leanings of his youth, Einstein expressed himself on the topic in words which reflect the metaphysical principle.

> "I have no better expression than 'religious' for this confidence in the rational nature of reality and in its being accessible, to some degree, to human reason. When this feeling is missing, science degenerates into mindless empiricism."

The 'Big Bang' Theory

Astute readers will have realised by now that the thesis promoted here is incompatible with the theory of the provenance of the universe as it is currently promoted. The Abbé Georges Lemaître conceived the idea, later mocked by British astronomer Fred Hoyle, as "based on the hypothesis that all the matter in the universe was created in one big bang..." The Abbé clearly had little exposure to St Thomas's thinking in his philosophical formation or else he forsook it in his enthusiasm for the categories of experimental science. Though I do not suggest that this was the Abbé's mind when he advanced the theory, there is behind its current promotion the materialist need to provide an explanation for reality which does away, or appears to do away, with the need for a Creator. Science will accept anything but an extrinsic (*efficient*) cause!

As I remarked in Chapter 1 scientists labouring under materialism's conventions are preoccupied with continua and temporal considerations. The problem with the 'big bang' theory ontologically is this. Before a 'big bang' could occur there must have been something to 'go bang'. *Ergo*, that *something* must have pre-existed it. But prior even to this there must have existed *a place* where that something could explode. Thus, if it did occur, the 'big bang' was not the first; at best it was only third, in a series of events that laid the foundations of the universe. Accordingly, the ontological order is—

Somewhere

Something

Explosion

117

The 'big bang' theory has a series of problems stemming from material-ism's simplistic grasp of reality. The first, proceeding on the impossible premise of non-being-somehow-existing, assumes that something can occur in a sea of nothingness. The second puts the cart before the horse; assumes that the explosion ('the bang')—which is an accident in meta-physical terms—could occur in the absence of a substance to support it. The third involves the gratuitous assertion that a substance, or substances, appeared spontaneously as a consequence of the explosion.

Let's look at the three ontological steps.

Somewhere is provided by Aristotle's heavenly body, *aether*, container of all other bodies. *Something* is provided by a pre-existing body of common material being. There might then have been the *Explosion*. If it did occur it may account for the elaboration, after aeons of time, of the celestial bodies we know today, the stars (suns), the planets, the moons, asteroids, comets, their parts, and the 118 elements of which they are comprised. But neither the heavenly body nor common mate-rial being brought itself into existence. To assert, or imply, that either did so is as fatuous as the other materialist premises cited.

It is reasonable to believe the two material bodies were created *ex nihilo* by Almighty God as it is reasonable to believe that He revealed how he created them and the ontological order in which He did so. In the very first words of *Genesis,* the first book of the Bible, we read:

In principio creavit Deus caelum et terram.

The Centre of the Universe

Science is loud in its condemnation of the cosmology of ages past which placed the Earth as the centre of the universe. It regards this as simplis-tic and coupled with the naiveties it associates with religion, chiefly the Catholic religion. Its votaries can demonstrate how peripheral to the almost infinite immensity of the universe is our own solar system and how insignificant, in such immensity, is our planet. Douglas Adams remarked this insignificance in humorous fashion in his fictional work *The Hitch-hiker's Guide to the Galaxy*:

"Far out in the uncharted backwaters of the unfashionable end of the western spiral arm of the galaxy lies a small unregarded yellow sun..."

But rejection of the ancient cosmology is simply another instance of modern science endorsing the shallow protocols of materialism and its attendant prejudice in favour of atheism. The ancient thinkers were more rational than our modern scientists.

In July 2015, 9 ½ years after it was launched, the probe *New Horizons* reached the furthest planet in the solar system *Pluto* and its chief moon *Charon*. I was teaching home-schooled children philosophy at the time and I invited them to consider the question: "Where do we go to now? What's our next space destination?" I drew to their attention that the nearest star, *Alpha Centauri*, is some 4.3 light years distant, had them calculate this distance in kilometres and work out how long it would take a hypothetical space ship travelling at 22 kilometres per second—the fastest at which we have so far been able to drive a space vehicle—for it to reach the star. The results are sobering.

The light emitted by the star takes 135,604,800 (60 x 60 x 24 x 365 x 4.3) seconds to reach us. Light travels at some 300,000 km/s. Therefore its distance from Earth is some 40,681,440,000,000 km. Dividing this by 22 gives the time for the trip to the star as 1,849,156,363,636 seconds, or 513,654,545 hours, or 21,402,273 days, or 58,636 years! Where in space do we go now? The answer is: nowhere outside our own solar system!

In another lesson I asked the students: "If God had meant man to fly what would he have given him?" There was the usual humorous response—"Wings!"—but I insisted: "Man does fly. So what is it that God has given him that enables him to do so?" The answer to that question assists in answering the question of the location of the centre of the universe.

What is it that God has given him that enables man to fly? *Intellect.* What is intellect? It is an immaterial power that enables its possessors not just to know singular things but to know their very natures. Aristotle did not underestimate its significance when he said: "The least degree of intellect in one is greater than the whole of the rest of material being", and again, "the one who possesses intellect is in a sense all things". It is intellect that frees us from the limitations of matter *because it is itself immaterial.*

One who has intellect does immaterial acts and, since do follows be, he can do so *because he is himself immaterial*. And, since only material things corrupt, the corruption of a man's body in death does not extend to the soul which gives him his essence and which is the seat of his intellect. What follows? The corruption of his body will not bring about his annihilation: when a man's body dies his soul endures. We did not make ourselves: we were made. And we were made not for this limited material existence only: we were made for eternity. That issue and its implications are, of course, beyond the scope of this book.

Where is intellect, this supreme power, to be found in the universe? *Here, on this planet and nowhere else!* The ancient thinkers were right: this Earth *is* the centre of the universe and the materialist endeavour to try and locate a *physical* centre is to no purpose. Moreover, since its ambit is beyond our knowledge, any attempt to do so is impossible.

The Value of Newton's and Einstein's Conceptualisations

1. The expression 'gravitational field' in modern scientific theory, whether used according to the Newtonian or the Einsteinian view, is compatible with a metaphysical view of the universe provided it is understood in a contrasting fashion to the expression 'magnetic field' when used of a body of iron. A celestial body is *not* the source of gravity as a body of iron is the source of the magnetism that surrounds it. The celestial body is but the focus of *aethereal* action—matter to *aether's* form, potency to *aether's* act—and to this extent its surroundings could be said to be part of the 'field' of *aether's* action about the body.

The celestial body's function is *be-determined. Aether's* function, in contrast, is *determine*. That a celestial body appears to be the source of a 'gravitational field' arises from the fact that the *action* of the *aethereal* matrix upon it is specified by the body's mass. The greater the mass, the greater the force *aether* exercises in investing it with circular motion—turning it aside constantly from rectilinear motion (*Newton's First Law*). And the greater the acceleration with which it is rotated, the greater the force exercised for the same mass—*Newton's Second Law*, $F = m\,a$.

2. The 'lensing' of a light ray around a massive body *may be conceived* as the effect of the mass of the body 'warping the space around it'

following the body's geodesic, the line of shortest distance over its spherical surface. But this is no more than mental being proposed by those who, because they are ignorant of the demands of the doctrine of causality, lack a true grasp of reality. There can be no effect, even in the furthest reaches of the universe, without an *extrinsic* cause acting.

If the thesis be accepted that has been advanced here, it is the universal agent, *aether,* investing the celestial body with circular motion which alters the direction of a light ray passing in proximity to it as an incident of the rotational force it exercises.

3. Time is the measure of change or movement. It is primarily mental being (the mind counting) yet it is mental being based in the real, for the mind counts real change, real movement. Metaphysically there is no reason why time noted by one observer should be identical with that noted by another. Newton opined that time was absolute. Einstein seemed to show it was not. What was absolute for both, did they but realise it, was the substance which underlies all reality, the heavenly body, *aether.* Newton's opinion appeals to it implicitly. But Einstein's does so too with his insistence on the fixity of c, 'the speed of light', for c is a property not of light but of light's proper substance, which is the heavenly body, *aether.* Beneath whatever relativity that may appear this principle of fixity abides.

4. Einstein's *General Theory of Relativity* predicts that a sufficiently compact mass of matter can deform 'space-time' to produce a 'black hole'. I suggest this may be explained in metaphysical terms by the dominance of *aether's* function as *ordinator* (orderer of the universe). As such it invests a particular celestial body with circular motion with such intensity that this suppresses its function as *lucifer* so as to impede it completely.

Science tells us that atomic clocks at different distances from the Earth's surface keep different times. A clock on the surface of the planet (i.e., closer to its centre) runs more slowly than one further away, as e.g., one on the top of a high mountain, and slower still than one on a GPS satellite in stationary orbit. The clock that is closer to the gravitational mass, 'deeper in its gravity well', is more affected by that mass. Each clock is determined in its operations by atomic resonance which has its foundation in the fixity of the *aethereal* matrix. The clocks differ in their readings yet disclose no defects in their operation. What occurs,

I suggest, reflects what befalls the light rays 'lensed' (or 'refracted') around those distant stellar bodies. There is no compromise of the 'speed of light', c, but *aether's ordering* function supervenes over its energy carrying function to delay the recording of an event, even by milliseconds. The alteration in the various clocks' readings marks the differing intensities with which *aether's* force is exercised in proximity to the centre of mass of the celestial body. This is not to say that the alterations are effected in a fashion identical (i.e., univocally) with that which obtains with 'lensing' around distant stars, but in a manner analogous. (See Chapter 1 for the meaning of 'analogous'.)

Science's current gravitational theory posits no causative influence in the circular movement of heavenly bodies. It can offer no explanation for the velocity of their rotation. Behind this insouciance is materialism's protocol of disregarding realities it cannot explain. Aristotle, in contrast, remarks circular motion as the primary *indicium* of *aether's* causative action.

Consistent with Aristotle's view I contend that *aether* is the instrumental *efficient* cause of the motion of the heavens. The *material* cause is the celestial bodies themselves. The *formal* cause is the *accidental* realities induced in them (circular motion, centripetal force, sphericity). The *final* cause is the order in the universe (ordination and subordination for the good of the whole). The principal *efficient* cause of the motions of the universe is Aristotle's intellectual being of immense power who, as St Thomas notes, is the Author of the creation and conservation of the whole of material being, Almighty God.

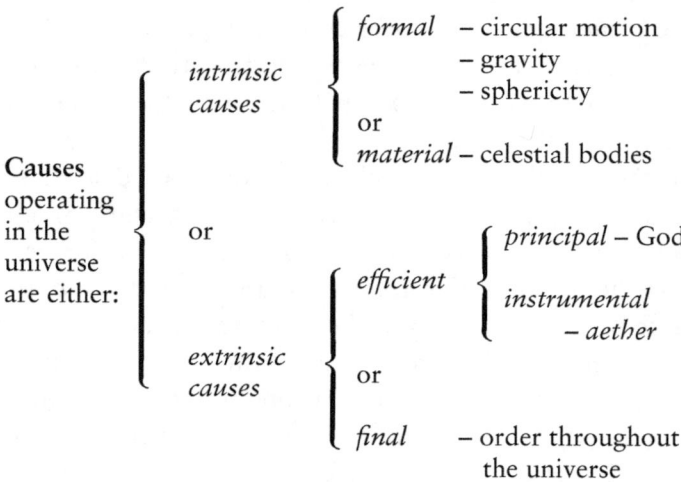

Man's place in the universe is a rational one. We are not, as materialists (no matter what their colour) opine, the mere result of the working
out of blind forces, but the pre-eminent beings in the universe because
we possess immateriality. We do immaterial acts, knowing not merely
that things are but *what they are*—their very natures.

The issue of our place in creation and the rational consequences to
be drawn from this realisation must abide another study. It is sufficient
here to insist that reason demands that the universe has a Creator and
Conserver who is intellectual. More than this: this being is an intellect
who is *Intellect Its Very Self*. (cf. Aristotle, *Metaphysics* XII, Ch. 9,
1074b 15 et seq.)

In 1865 in his work *The Temporal Mission of the Holy Ghost*, the
Catholic Archbishop of Westminster, Henry Edward Cardinal Manning,
summarised the issue for mankind in two of three points he made,
which I commend the reader to ponder:

> "It is a violation of reason not to believe in the existence of God...
> it is a violation of our moral sense not to believe that God has
> made himself known to man..."

It is fitting, then, to close this work with a passage the Catholic Church
insists comes from the revelation that God has made of Himself to all
mankind:

> "With heaven my throne and earth my footstool, what house
> could you build me, what place could you make for my rest? All
> of this was made by my hand; all of this is mine—it is the Lord
> who speaks. But my eyes are drawn to the man of humble and
> contrite spirit who trembles at my word."
>
> *Isaiah* 66: 1-2

GLOSSARY

Like modern experimental science, the science of metaphysics has its own terminology. Its understandings (concepts) are radically different from those which, under the influence of the thinkers of the Enlightenment, flourish today. If the reader is to follow the arguments in the text he must have some grasp of this mode of thinking. It was, until the sixteenth century, the mode of thought accepted universally by men.

Accident	A reality that exists only in some substance. Accident = be-in-other. There are nine of them: quantity, quality, relation, when, where, action, passion, habitus, situs.
Act	'Does-be-ness': this may seem a facile way to express it but it is the meaning of the word in Latin, *ac-tus*. Contrast with potency.
Action	Accident, the exercise of the causality of an efficient cause. Gravitational force is an example.
Analogical (analogy)	(Logical term) A predicate, signifying a character found in different entities which is some-wise same and some-wise un-same, and more un-same than same; as, e.g., the term 'healthy' said of food, of the climate and of normal human constitution; or 'being' said of a colour or a sound (accidents) or a horse or a man (substances).
Categories	The ten ways in which a material being exists identified by Aristotle in two classes, substance and (nine different) accidents.

Concept	(Logical term) A sign generated by the intellect to signify something real or not real. Contrast 'word'. "Concepts are the material of which our mental acts, true and false, consist." (Joyce, *Principles of Logic*, 1916, Ch. II)
Contrary	Contrary opposition is opposition between two classes which are furthest removed from each other among those which belong to the same genus.
Contradictory	Contradictory opposition is the opposition between a term and its negation.
Efficient Cause	Causes by acting, by making or doing.
Equivocal (equivocity)	(Logical term) A predicate, signifying a character found in different entitles which is simply un-same, as when we say 'cricket' of the insect and the game, or 'board' said of a wooden plank and a group of people. Contrast 'univocal' and 'analogical'
Final Cause	Causes by being desired: the end of one's action. It is the first cause in intention and the last in execution. It begins the process of causation and ends it when the end intended is achieved.
Form	The determining influence in any thing; the influence that makes a thing be that thing. Itself not material, when it is blended with matter the result is the material thing.
Formal Cause	Causes by determining
Habitus	An accident indicating covering or clothing.
Intellect	The power proper to man, the rational animal, by which he knows the natures of things. The proper formal object of the

human intellect is the *quiddity* (the 'what-ness') of material things. Man knows not only *that* things are but *what* things are.

Inverse square law	The strength of a specified physical force is inversely proportional to the square of the distance from the source.
Knowledge	To have something in self formally and not materially. Knowledge is twofold. *Sensitive knowledge* is of singulars only. *Intellective knowledge* is of their universal natures. The brute animal knows *that* a thing is: man knows *what* the thing is.
Logic	The science which directs the operations of the mind in the attainment of truth. "The object of Logic is… the thing considered as an object of thought endowed with attributes of the conceptual order." (Cardinal Mercer quoted in Joyce, *Logic*, note to ch. 1)
Materialism	A philosophy that holds there is only one cause, matter.
Matter	Material potency, 'can-be-ness'. It is divided into *prime* matter and *secondary* matter. Prime matter is never found, never seen, alone for it only ever exists in compound with some form. It is unknowable in itself for we know a thing via its formality and prime matter has no formality. Prime matter is pure potency. *Secondary* matter is matter already formed, as found, e.g., in one or other of the 118 elements or air, water, iron, stone etc.
Material Cause	Causes by being determined
Metaphysics	Title of a text by Aristotle written after the *Physics*. The word means 'after, or beyond, the physics'. The text deals not with any particular form of being but with being

simpliciter. The term "sometimes stands for (Aristotelian) philosophy in general, sometimes... for that part of philosophy known as Ontology (the philosophy of being as being)". (cf. Joyce, *Logic*, ch. 1) "The object of metaphysics is being considered in abstraction from all individual determinations and material properties..." (Joyce: Note to ch. 1, quoting Cardinal Mercier)

Nature	The order and elements established by the author of all being. The root 'na-' means 'that which is given', hence *na*ked, *na*tive, *na*tural, in*na*te.
Ontological	The order of reality. Contrast 'temporal'. Things may happen together in time but in the order of reality one may precede the other.
Passion	Accident, the reception of the effect and action of the efficient cause.
Potency	'Can-be-ness': this may seem facile in expression but it reflects the meaning in the Latin noun *pot-en-tia*. It is real, not imaginary, being. Contrast with act.
Quantity	The first accident, the one that gives extension and parts, a body, to a material substance.
Quality	The second accident and the one most crucial to the material substance after quantity. It forms and qualifies, determines 'of what sort' (*qualitatis*) the thing is, in a variety of ways.
Relation	Accident whose whole reality is be-towards another.
Situs	Accident revealing posture as, e.g., sitting, lying, up-side-down.

Substance	A reality that exists in itself. Substance = be-in-self. It is that which 'stands under' the accidents that befall it.
Subjectivism	A philosophy grounded in the observer rather than in the thing, in the subjective rather than the objective. Owes its provenance since the 17th century to the thinking of René Descartes.
Temporal	The order of time. Contrast 'ontological'.
Thing	An existing entity. By transference it may refer to what exists in mind as well as in the real, or in mind without existing in the real.
Think	To 'become the thing', a facility open only to those possessed of the power of intellect. (Brute animals cannot think.)
Univocal (univocity)	(Logical term) A predicate, signifying a character found in different entities simply the same as when we call 'dog' the individuals Fido, Rex, Shep, etc. Contrast 'equivocal' and 'analogical'.
When	Accident dictating the point in time in which the material substance exists. Time is the measure of change. It consists formally in the mind measuring, materially of the reality the mind is considering.
Where	Accident dictating the place in which the material substance exists.
Word	(Logical term) A Vocal or written sign of an understanding, or concept, which is a sign, in turn, of something real or not real. Contrast 'concept'.

BIBLIOGRAPHY

Adler, Mortimer J, *How to think about God: a guide for the 20th Century Pagan*, New York (Simon & Schuster), 1980

Aquinas, St Thomas, *In De Anima, In Metaphysica, In Physica, De Sensu, In II Sententiae, Summa Theologiae, De Veritate*

Aristotle, *De Anima, De Caelo, De Generatione et Corruptione, Physica, Metaphysica*

Boland, D. G, LL.B, Ph.D, *God and the Theory of Everything*, available at https://www.superflumina.org/ PDF_files/d-boland-god-the-theory-of-everything.pdf

Decaen, Christopher A, Ph. D., *Aristotle's Aether and Contemporary Science*, The Thomist 68 n.3, July 2004

Eicher, David J., *Comets, Visitors from Deep Space*, New York (CUP), 2013

Eliot T. S., *The Rock*, New York, 1934, *Four Quartets*, New York, 1943

Foster, K, and Humphries, S, *St Thomas's Commentary on Aristotle's De Anima*, Notre Dame, Indiana, 1994

Jeffrey, G. B. and Perrett W., *Sidelights on Relativity*, New York (Dover), 1983

John of St Thomas, *Cursus Philosophicus*

Joyce, G. H., *Principles of Logic*, London, 1916

Kavanaugh, Kieran, O.C.D. and Rodriguez, Otilio, O.C.D., *Collected Works of St John of the Cross*, Washington D.C., 1979

Massey, Steve, *Exploring the Moon*, Sydney, 2004

Maxwell, James Clerk, *A Dynamic Theory of the Electromagnetic Field*, 1864

Newton, Sir Isaac, *Philosophiae Naturalis Principia Mathematica* (1687), *Letters to Dr Richard Bentley* (1756)

Sayers, Dorothy L, *The Unpleasantness at the Bellona Club*, London, 1921

Sire, H.J.A., *Phoenix from the Ashes—The Making, Unmaking and Restoration of Catholic Tradition*, Kettering OH (Angelico Press), 2015

Superflumina.org, (the author's website) *Pity the Poor Atheist*, https://www.superflumina.org/PDF_files/pity_theatheist.pdf;

Science and Aristotle's Aether, http://www.superflumina.org/PDF_files/aether_science.pdf

University of California, Irvine (uci.edu)

Woodbury, Fr A. M., S.M., Ph.D., S.T.D., *Ontology; General Natural Philosophy & Cosmology*, available through https://austinwoodbury.com/pg-search.php

Wikipedia—http://en.wikipedia.org – topics: *Barycentre, Mass, Tides, Forces*

INDEX

Note: 'The Philosopher', when used in the text by St Thomas, is a reference to Aristotle. References to the Glossary are shown with the letter 'G'.

ABOUT THE AUTHOR

The author spent some 35 years practising as barrister and solicitor in New South Wales. His studies at Sydney University's Law School in the 1960s were balanced with studies in the philosophy of Aristotle and Aquinas at Sydney's Aquinas Academy.

Philosophy has ever been his first love. His authority to offer the commentary and criticism on the philosophical issues in the text derives from his studying under teachers Fr Austin M Woodbury S.M., Ph.D., S.T.D., foremost philosopher and theologian of the Catholic Church in Australia and his assistants, John Ziegler B. Sc., Geoffrey Deegan B.A., Ph.D. and Donald Boland Ll.B, Ph.D.